A FORMULARY OF CANDY PRODUCTS

By

Norman Harris
U.S. Army Natick Research, Development & Engineering Center

Dr. Martin S. Peterson
(Deceased)
Editorial Consultant

Dr. Silvio Crespo
Vice President—Technical Emeritus, Wilbur Chocolate Co.

CHEMICAL PUBLISHING COMPANY, INC.
New York 1991

Chemical Publishing Co., Inc.

ISBN 0-8206-0333-3

Printed in the United States of America

CONTENTS

PREFACE

Formulas for confections shown here contain the basic ingredients needed to produce each product. Most candies illustrated *do not* use food additives other than the main constituents needed for preservation. For prolonged shelf life other additives or adjuncts may be essential such as antioxidants, emulsifiers to control texture (maintain freshness), and fats containing higher melting fractions. To control cost the use of milk replacers such as whey with artificial milk flavors could be substituted.

The creative role of transforming a candy into the delicacy sought by the most discriminating consumer is up to the most competent technologist. This means developing the right combination of ingredients, processing them with extra care, and making sure that the product is properly protected thereafter until consumption. Most candy contains sugar in an aqueous supersaturated state. Candy is sweet in taste and its texture can vary from very soft to hard, crispy, chewy and grainy depending on the ingredient and processing factors. Candy probably ranks as the most calorie dense food. It can be stored without refrigeration and was used in the sixties to stock shelters for an emergency. People use it worldwide for quick energy and it is used as a dessert item in operational rations for the Armed Forces worldwide.

The mention of one brand or ingredient in a formula contained in tis book should not be construed to mean that one of similar quality is not equivalent or superior in function. Moreover, the user of these ingredients is responsible for checking for any patent infringement should the latter exist. Any suggestions, criticisms, or other comments from users of this book concerning any formulae will be welcomed.

Norman E. Harris
Martin S. Peterson (deceased)
Silvio Crespo

ACKNOWLEDGEMENT

This collection of industrial confections was made possible by the following companies:

Amstar Sugar Co.
Archer Daniels Midland (ADM)
California Almond Growers Exchange
Central Soya Chemurgy Division
C & H Sugar Co.
CPC International, Inc.
Crompton & Knowles
A/S Kogenhavns Pektinfabrik
Kraft General Foods Corp.
Penwest Ltd.
Pillsbury Co.
RJR Nabisco
Savage Brothers Co.
Tate & Lyle PLC
Van Den Bergh Foods
Wilbur Chocolate Co.

A special thank you should go to my son Chris J. Harris, who programmed my hand computer for calculation of the ingredient percentages in each of the formulas. In addition, Ms. Irene Ash deserves exceptional praise for her diligence and patience during word processing, printing, and critical suggestions to help complete this task.

Norman E. Harris

ABBREVIATIONS

Bé .... Baumé
c .... cup
cc .... cubic centimeters
D.E. .... dextrose equivalent
F .... ° Fahrenheit
g .... grams
gal .... gallon
h .... hour
in. .... inch
lb .... pound
meq .... milliequivalents
m.p. .... melting point
min .... minutes
oz .... ounce
pbw .... parts by weight
pt .... pint
qt .... quart
rpm .... revolutions per minute
Tbl .... tablespoonful
tsp .... teaspoonful
wt. .... weight

Chapter 1

AERATED CANDIES

Aeration is employed in candy production to bring about desirable changes in texture and consistency. Aeration produces effects that cannot be duplicated by other means. Aerated candies are more creamy, more fragile, less chewy, and less likely to cling to the teeth than other types of candy.

Fondants Used in Making Aerated Candies

Fondant

No. 1

(Basic Fondant)

			% by Wt.
Granulated sugar	80	lb	73.0
Corn syrup (43 D.E.)	20	lb	18.0
Invert sugar	10	lb	9.0
			Total 100.0

Procedure:

Boil ingredients to 242 F (sea level), then cool and cream into fondant.

Source: Crompton & Knowles

No. 2

(For Cream Wafers and Cream Dipped Bonbons)

			% by Wt.
Refined cane sugar	50	lb	77.89
Water	12	lb	18.69
Nulomoline (invert sugar)	2	lb	3.12
Gelatin (dissolved in hot water)	1	oz	0.09
Water (hot)	2	oz	0.21
			Total 100.00

Procedure:

Place the sugar, water, and **Nulomoline** into a clean kettle. Turn on heat and stir batch occasionally until it reaches boiling point. Carefully wash down all grains of sugar that adhere to sides of kettle, and boil the batch as rapidly as possible to 244–246 F. Turn off heat and immediately add the dissolved gelatin. Stir the batch gently to mix gelatin, then promptly pour the batch onto a cold marble or metal slab which has been lightly dampened with water. Now take 2 oz of cold water and sprinkle the surface of batch using a brush for this purpose.

Allow the batch to remain undisturbed until it has cooled to approximately 90 F, then work into fondant. Allow the fondant to set about $^1/_2$ h, then take a portion of the fondant, place it into a double boiler and stir continuously until it has melted and registers a temperature of 130–135 F, then use for bonbon work.

For cream wafers, it is suggested that the fondant be heated to approximately 145 F and then colored and flavored and cast on waxed paper or rubber mats.

When the wafers are cold, turn them upside down to thoroughly dry the bottom of the wafer, then place them on trays or pack. (For each 550 ft above sea level, reduce boiling temperature by 1 F).

Source: Crompton & Knowles

No. 3

(For Dipping Purposes)

			% by Wt.
Granulated sugar	85	lb	66.79
Corn syrup	15	lb	11.79

Nulomoline (invert sugar)	7	lb	5.50
Water	20	lb	15.72
Hard coconut butter (80–90 F m.p.)	2	lb	0.10
Medium grade gelatin (soaked in 5 oz cold water)	2	oz	0.10
			Total 100.00

Procedure:

Place the water, sugar, **Nulomoline**, and corn syrup into a kettle and bring the batch to the boiling point. Stir occasionally so as to be certain that the sugar is dissolved when the batch begins to boil. Wash down the sides of the kettle to remove all grains of sugar. Place a thermometer into the batch and cook to 244–248 F.

Turn off the heat, add the gelatin solution and mix the batch gently until the gelatin is dissolved. Cool the batch to approximately 105 F, add the coconut butter which should be melted, then beat into fondant. It is advisable to use the fondant the same day on which it is made.

If this fondant is to be used in a coating machine, it is suggested that it be boiled to 238 F, then cooled to 100 F and beaten in the usual manner.

When preparing the fondant for machine work, it is suggested than an additional 5 lb of **Nulomoline** be added to 100 lb of fondant which should be heated to 105–110 F.

Source: Crompton & Knowles

Aerated Candies

Almond Crystallized Creams

			% by Wt.
Casting fondant	140	lb	47.62
Granulated sugar	100	lb	34.01
Corn syrup	10	lb	3.40
Water	3	gal	8.50
Nonpariel Supreme almonds (18/20 or 20/22 count)	15	lb	5.10
Frappé whip (pink color, raspberry flavor)	4	lb	1.37
			Total 100.00

Procedure:

Place casting fondant into cream melting kettle, heat to 115–120 F. Meanwhile heat sugar, corn syrup, and water to boiling. Cook to 242–244 F, then add to the fondant, mixing thoroughly. Add the color, flavor, and frappé. Cast into starch impressions or rubber molds containing whole almonds. When creams have set, remove from starch. Prepare crystallizing syrup as follows: Cook 300 lb granulated sugar and 15 gal water. Boil and stir until batch registers 33.5 Bé at boiling point. If a Baumé gauge is not available, boil to 221 F at sea level. Cool to 80–90 F, then ladle syrup over creams in crystallizing pans or immerse wire baskets containing creams in cool syrup. Permit creams to remain in syrup 7–10 h, drain and turn creams onto waxed paper.

Source: California Almond Growers Exchange

Almond Paste Centers

(Coat with Chocolate or Crystallize)

		% by Wt.
Almond paste	10 lb	30.27
Short bonbon fondant	10 lb	30.27
Nulomoline (invert sugar)	1 lb	3.04
Fine powdered sugar	12 lb	36.33
Sucrovert (liquid invertase)	0.5 oz	0.09
Vanilla, rum, or pistachio flavor		to suit
Color		to suit
		Total 100.00

Procedure:

Place all of the above ingredients into an upright mixer or knead by hand to form a firm paste or dough. Increase or decrease the percentage of powdered sugar to obtain a paste sufficiently firm to roll by hand into pieces of the desired size and shape.

The almond paste centers may be roughly formed into pieces of the desired size, then placed into rubber molds. The formed centers are then removed from the molds and allowed to dry on the surface and are later crystallized or coated with chocolate.

The almond paste center mass may be formed into a sheet of varied colored layers, then cut into squares, oblongs, and triangles, then later crystallized or coated.

The almond paste center mass may also be rolled into a sheet and cut into various shapes using a standard hand lozenger cutter, and the centers may then be coated with fondant bonbon coating.

Source: Crompton & Knowles

Almond Whip Bar

			% by Wt.
Milk chocolate coating	10	lb	36.95
Hard coconut butter	5	lb	18.48
Frappé	7	lb	25.86
Buttered, diced roasted almonds*	5	lb	18.48
Salt	1	oz	0.23
Rum flavor			to suit
			Total 100.00

Procedure:

Mix melted coconut butter and melted chocolate together. Allow to cool to approximately 82 F. Pour in small kettle cake mixer. Whip, adding frappé. Beat or whip at second speed until fluffy. Add salt and flavor and 5 lb almonds. Pour on waxed paper and spread approximately $^{1}/_{2}$ in. thick. When batch has set and cooled, remove the waxed paper and cut to desired bar size. Dip in milk or dark chocolate coating. May be topped with bits of blanched almonds if desired.

Source: California Almond Growers Exchange

Assorted Cream Squares

			% by Wt.
Granulated sugar	20	lb	47.62
Corn syrup	10	lb	23.81
Frappé No. 1 (q.v.)	4	lb	9.52

* If you prefer, use select sheller run. Roast or chop or crush to suit.

Powdered or macaroon coconut	2	lb	4.76
Basic Fondant (q.v.)	6	lb	14.29
Salt, flavor, color			to suit
			Total 100.00

Procedure:

Boil the sugar and corn syrup with sufficient water to 242 F (sea level). Set the batch off the fire, add the Frappé, coconut, and basic fondant, and mix well, then add salt, flavor, color, etc. and mix thoroughly. Divide the batch into two or more sections:

Section No. 1: White color, vanilla flavor; coarse cut citron mixed together.

Section No. 2: Yellow color, **Nulco MeloBan** (powdered ripe banana) flavor, plus sliced Brazils.

Section No. 3: Pink color, Roman punch flavor, plus glaced cherries cut in half.

The fruit and nuts are to be worked into the surface of the batch only. Spread about $^5/_8$ in. thick.

FRAPPÉ NO. 1:

Corn syrup	50	lb
Invert sugar	50	lb
Albumen	1	lb
Water	2	lb

Procedure:

Cook 50 lb of corn syrup to 245 F (sea level), shut off the heat and add 50 lb of invert sugar and stir until it is melted. Place the batch into a marshmallow beater, start the beater, then gradually add 1 lb of albumen which has been previously dissolved in 2 lb of water and beat the batch until light.

BASIC FONDANT:

Granulated sugar	80	lb
Corn syrup	20	lb
Invert sugar	10	lb
Water		

Procedure:

Boil 80 lb of granulated sugar, 20 lb of corn syrup, 10 lb of invert sugar, plus sufficient water to dissolve the sugar, to 242 F (sea level), then cool and cream into fondant.

Source: Crompton & Knowles

Ban-O-Bar

		% by Wt.
Granulated sugar	85 lb	70.21
Nulomoline (invert sugar)	15 lb	12.39
Water (to dissolve the sugar)		
Sucrovert (invertase)	1 oz	0.05
Special Frappé (q.v.)	10 lb	8.26
Cut or shredded orange peel (preserved)	6 lb	4.96
Nulco MeloBan (mixed with $1^1/_2$ lb water)	3.5 lb	4.13
Yellow color		to suit
Citric acid		to suit
		Total 100.00

Procedure:

Boil sugar, **Nulomoline**, and water to 250 F. Then cool batch to approximately 125 F and beat into fondant.

While beating, add the **Sucrovert, Nulco MeloBan** (powdered ripe bananas) which has been mixed with water, and the color; and when the batch grains, add the Special Frappé and citric acid. When this is well mixed, work in the orange peel.

Permit the batch to cool to room temperature, then form into bars of the desired shape and coat with milk chocolate.

SPECIAL FRAPPÉ:

Sugar	15 lb
Corn syrup	15 lb
Water	
Nulomoline	20 lb
Albumen	2 lb
Water	4 lb

Procedure:

Boil 15 lb of sugar, 15 lb of corn syrup with sufficient water to dissolve the sugar to 250 F. Turn off the heat, add at once 20 lb of **Nulomoline** and stir until melted. Place the batch into a marshmallow beater, start the beater, then gradually add 2 lb of albumen which has been properly dissolved in 4 lb of water; beat the batch until quite light.

Source: Crompton & Knowles

Cast Cream Centers for Bars, Easter Eggs, etc.

			% by Wt.
Fondant (q.v.)	100	lb	60.23
BOB:			
Sugar	25	lb	15.06
Corn syrup	15	lb	9.03
Nulomoline (invert sugar)	10	lb	6.02
Water	6	lb	3.61
Standard Frappé (q.v.)	10	lb	6.03
Convertit (invertase)	0.5	oz	0.02
			Total 100.00

Procedure:

Place 100 lb of the fondant into a cream melting kettle and heat to approximately 120 F. Meanwhile prepare the bob.

Boil all together to 240 F. Pour this bob into the melted fondant mixing continuously, then add standard frappé. Mix well. Then add **Convertit**. Mix thoroughly, then add color and flavor. Test the batch so as to be certain it will register 150–155 F.

FONDANT:		
Sugar	75	lb
Corn syrup	25	lb

Procedure:

Cook to 245 F.

STANDARD FRAPPÉ:

Nulomoline (invert sugar)	50	lb
Corn syrup	50	lb
Albumen (dissolved in water)	1	lb
Water	2	lb

Procedure:

First cook 50 lb of corn syrup to a temperature of 245 F. Shut off the steam and add 50 lb of cold **Nulomoline** and stir until the latter is melted. Place the batch into a marshmallow beater, start the machine, then add the albumen solution. Beat until light and use as needed.

More or less frappé may be used according to the weight of cream center desired. (For each 550 ft above sea level, reduce the boiling temperature by 1 F.)

Source: Crompton & Knowles

Cast Cream Cordial Cherries

			% by Wt.
Basic Fondant (q.v.)	100	lb	51.73
BOB:			
Sugar	70	lb	36.21
Nulomoline (invert sugar)	5	lb	2.59
Water	18	lb	9.31
Sucrovert (invertase)	5	oz	0.16
			Total 100.00

Procedure:

Prepare basic fondant.

Boil bob altogether to 230 F.

The bob. when mixed with the fondant, registers a temperature of 165 F and is cooled to 160 F before adding **Sucrovert**.

Casting temperature 160 F.

(No frappé used in this formula.)

BASIC FONDANT:

Corn Syrup	80 lb
Nulomoline (invert sugar)	10 lb
Water (to dissolve the sugar)	

Procedure:

Cook altogether to 242 F.

CHERRY SPECIFICATIONS:

Count 900/1000 per U.S. gal. The syrup in which the cherries are packed contains 40.5% solids. The ratio of syrup to cherries is approximately 4 lb/ 1 gal. The density of the syrup in which the cherries are received is 40.5.

One ounce of tartaric acid is mixed with 1 gal of the cherries in syrup and allowed to set overnight.

One pound of cream center batch is used to cover 40 cherries by making two castings. Forty cherries when drained weigh approximately 6 oz. Part of the batch is colored orange and flavored with orange oil and cast into starch impressions and later used as yolks for eggs.

Source: Crompton & Knowles

Cast Whipped Cream for Easter Eggs and Bars

		% by Wt.
Basic fondant (q.v.)	100 lb	44.05
BOB:		
Sugar	50 lb	22.03
Corn syrup	25 lb	11.01
Water	12 lb	5.29
Frappé No. 3 (q.v.)	40 lb	17.62
		Total 100.00

Procedure:

Prepare basic fondant.

Boil bob all together to 240 F.

Pour the bob into the fondant, mix well, then add Frappé No. 3.

Casting temperature 150–155 F. Density of centers when coated is 82.6.

To make a whipped cream cast the batch into half egg shaped molds, then the cast cream egg yolk center, previously prepared, is dropped into the whipped cream and immediately another half of the whipped cream is placed on top.

BASIC FONDANT:

Granulated sugar	20 lb
Sweetose corn syrup	20 lb
Water (to dissolve the sugar)	

Procedure:

Boil altogether to 242 F, then cool and beat into fondant.

FRAPPÉ NO. 3:

Corn syrup	50 lb
Nulomoline	50 lb
Gelatin	1.5 lb
Water	3 lb
Albumen	1.5 lb
Water	3 lb

Procedure:

Boil corn syrup to 245 F. Shut off the steam, add **Nulomoline**, and stir until melted. Place the batch into a beater, start the beater, then add gelatin which has been soaked in 3 lb of water, and beat until dissolved. Then gradually add albumen which has been previously dissolved in 3 lb of water and beat the batch until it is light.

Source: Crompton & Knowles

Chocolate Butter Creams

		% by Wt.
Dairy butter (salted)	5 lb	7.14
Coating chocolate	7 lb	10.00
Fondant (q.v.)	50 lb	71.43
Invert sugar	3 lb	4.29
Frappé	5 lb	7.14
Flavor (vanilla or Roman punch)		to suit
		Total 100.00

Procedure:

Melt together the dairy butter with good coating chocolate. Then add 50 lb of good fondant which you ordinarily use in making cast cream centers. Stir together thoroughly. Add the invert sugar and frappé. Add vanilla flavor or Roman punch. Mix well and heat to approximately 135 F and cast into starch.

FONDANT:

Sugar	75 lb
Corn syrup	25 lb

Procedure:

Cook to 245 F.

Source: Crompton & Knowles

Chocolate Patties

Sugar	10 lb
Glucose (corn syrup)	2.5 lb
Water	1 qt
Chocolate liquor	2 lb
Vanilla bean or cinnamon oil	

Procedure:

Cook sugar and glucose (corn syrup) with water to 236 F, and before you start to work the batch, add good chocolate liquor and the seeds of one vanilla bean, or a drop or two of cinnamon oil. Great care should be exercised, for these wafers will spot more easily than any other color or flavor.

Do not heat the cream more than is absolutely necessary.

Source: Savage Brothers Co.

Chocolate Pudding Dessert

(To be Cast into Starch)

PART 1:

Corn syrup	10 lb
Condensed milk (skimmed or whole)	30 lb

Coconut oil (76 F m.p.)	5 lb
Nulomoline (invert sugar)	10 lb
Unsweetened chocolate	10 lb

PART 2:

Powdered salt	6 oz
Basic fondant (q.v.)	100 lb
Butter and vanilla flavor	to suit
Sucrovert	1 oz
Frappé No. 1 (q.v.)	15 lb

Procedure:

Place Part 1 altogether into a kettle and heat until the batch is fluid, then turn on the steam and cook the batch to a medium soft ball. Add 6 oz of powdered salt, then add 100 lb of basic fondant, mix well, add butter and vanilla flavor and 1 oz of **Sucrovert** and mix well. Now add 15 lb of Frappé No. 1, mix thoroughly, then heat the batch to 150–155 F and cast into starch impressions of the desired size and shape.

BASIC FONDANT:

Granulated sugar	80 lb
Corn syrup	20 lb
Invert sugar	10 lb
Water	

Procedure:

Boil granulated sugar, corn syrup, and invert sugar, plus sufficient water to dissolve the sugar, to 242 F (sea level), then cool and cream into fondant.

FRAPPÉ NO. 1:

Corn syrup	50 lb
Invert sugar	50 lb
Albumen	1 lb
Water	2 lb

Procedure:

Cook corn syrup to 245 F (sea level), shut off the heat and add invert sugar and stir until it is melted. Place the batch into a marshmallow beater, start the beater, then gradually add albumen which has been previously dissolved in water and beat the batch until light.

Source: Crompton & Knowles

Coconut Cream Bars and Eggs

(Cast into Starch)

			% by Wt.
PART 1—FONDANT:	75	lb	43.45
PART 2—BOB:	60	lb	34.76
PART 3—COCONUT:	10	lb	5.78
PART 4—STANDARD FRAPPÉ:	10	lb	5.78
Sucrovert	2	oz	0.07
Vanilla flavor			to suit
			Total 100.00

Procedure:

Pour the bob into 75 lb of the fondant (Part 1), mix well, then add 10 lb of Standard Frappé (Part 4) and mix well. Now add 2 oz of **Sucrovert** and while mixing add vanilla flavor and 10 lb of the freshened coconut (Part 3). Mix well and cast into starch. Casting temperature 150–155 F.

PART 1—FONDANT:

Sugar	80	lb
Corn syrup	10	lb
Nulomoline (invert sugar)	10	lb
Water	22	lb

Procedure:

Boil altogether to 242–244 F, then cool and beat into fondant in the usual manner.

PART 2—BOB:

Sugar	45 lb
Corn syrup	5 lb
Water	10 lb

Procedure:

Boil altogether to 236–240 F.

PART 3—COCONUT:

Water	2.5 lb
Nulomoline (invert sugar)	5 lb
Fine desiccated coconut	20 lb

Procedure:

Heat the **Nulomoline** and water to approximately 150 F, add the coconut, mix well and allow to set for half an hour before adding to the batch.

PART 4—STANDARD FRAPPÉ:

Corn syrup	50 lb
Nulomoline (cold)	50 lb
Albumen	1 lb
Water	2 lb

Procedure:

Boil the corn syrup to 245 F. Shut off the steam and add the **Nulomoline** (cold) and stir until it is melted. Place the batch into a marshmallow beater, start the beater, then add the albumen which has been previously dissolved in water and beat until light in texture.

Source: Crompton & Knowles

Conserve Mints

(Cast into Rubber Molds or Starch)

		% by Wt.
Granulated sugar	40 lb	55.17
Water	10 lb	13.79

Nulomoline (invert sugar)	2.5 lb	3.45
Powdered sugar	20 lb	27.59
Peppermint oil		to suit
Color		to suit
		Total 100.00

Procedure:

Heat the water, sugar, and **Nulomoline** to the boiling point, wash down all sugar crystals from the side of the kettle, and cook the batch to 234–237 F (at sea level).

Immediately add $^1/_2$ of the powdered sugar (previously sifted), mix well, add the mint flavor, then add the remainder of the sifted powdered sugar, mix rapidly until smooth, allow the batch to cool to 155 F, then cast into rubber molds which have been tempered to approximately 75–80 F. When the candies are sufficiently firm, remove them at once from the rubber molds.

If desired, the conserve mints may be crystallized in a cool crystallizing syrup of 33–33.5 Bé; or the conserve mints may be sprayed with a solution of candy coloring using an atomizer; or the color may be brushed over the surface of the mints.

Source: Crompton & Knowles

Cordial Fruit Cream Centers

Cream Fondant Base (q.v.)	5 lb
Convertit	1 tsp

Procedure:

Melt 5 lb of Cream Fondant Base (q.v.) heating to 160 F. Add 1 tsp. of **Convertit** (invertase) and color and flavor, if desired.

Meanwhile, drain cherries, kumquats, or pineapple cubes from the syrup for 10–15 min.

Dip the fruit in the fondant and when the fondant coating has set, rub a little chocolate on base of each piece; then coat with chocolate.

When coating the cordial fruit cream centers with chocolate, place a small spot of coating, about the size of a dime, on dipping paper. Then place the freshly coated center on top of the chocolate spot, thus producing a thick coating on the bottom of the candy to check leaking.

CREAM FONDANT BASE:

		% by Wt.
Granulated sugar	8 lb	64.0
Corn syrup	1 lb	8.0
Nulomoline (invert sugar)	1 lb	8.0
Water	2.5 lb	20.0
		Total 100.0

Procedure:

Heat all together to boiling point. Wash down all crystals of sugar that adhere to pan and boil the batch to 240–242 F (sea level). Cool syrup to room temperature and beat into fondant.

Place the fondant into covered containers for future use, rubbing a little **Nulomoline** over the surface of the fondant to prevent surface drying.

Source: Crompton & Knowles

Coronation Bar

(For Bar, Bulk, or Package Goods)

		% by Wt.
Part 1 (q.v.)	65 lb	36.09
Part 2—Basic Fondant (q.v.)	100 lb	55.54
Vanilla and Butter Flavors (q.v.)		8.37
		Total 100.00

Procedure:

Add the 100 lb of Basic Fondant to Part 1 and mix thoroughly, then add the flavors and **Sucrovert** (invertase) and mix well. Now work into the batch the Frappé No. 1 and mix thoroughly.

Casting temperature 155–160 F.

Cast the batch into large impressions into which have first been placed yellow-colored cream centers. Fill the impressions and when the centers have set, remove them from the starch and coat with light or dark chocolate.

PART 1:

Corn syrup	10 lb
Granulated sugar	5 lb

Condensed milk (skimmed or whole)	25	lb
Coconut oil (76 F m.p.)	5	lb
Nulomoline (invert sugar)	10	lb
Unsweetened chocolate	10	lb

Procedure:

Place all together into a kettle and heat and stir the batch constantly until it boils; then add 6 oz of powdered salt and turn off the heat.

PART 2—BASIC FONDANT:

Sugar	80	lb
Corn syrup	20	lb
Nulomoline (invert sugar)	10	lb
Water (to dissolve sugar)		

Procedure:

Cook all together to 242 F, then cool and beat into fondant.

VANILLA AND BUTTER FLAVORS:

Sucrovert (invertase)	1	oz
Frappé No. 1 (q.v.)	15	lb

FRAPPÉ NO. 1:

Nulomoline (invert sugar)	50	lb
Corn syrup	50	lb
Albumen (dissolved in water)	1	lb
Water	2	lb

Procedure:

Cook the corn syrup to 245 F. Shut off the steam and add the **Nulomoline** (cold) and stir until the latter is melted. Place the batch into a marshmallow beater, start the beater, then gradually add the albumen solution and beat until light.

Source: Crompton & Knowles

Cream Cuts

			% by Wt.
PART 1:			
Granulated sugar	50	lb	36.45
Corn syrup	25	lb	18.22
Water (to dissolve the sugar)			
PART 2:			
Desiccated coconut	10	lb	7.29
Nulomoline (invert sugar)	3	lb	2.19
Water	2	lb	1.46
PART 3:			
Basic Fondant (q.v.)	35	lb	25.51
Frappé No. 1 (q.v.)	10	lb	7.29
Coconut butter	2	lb	1.46
Salt	3	oz	0.13
			Total 100.00

Procedure:

PART 1: Boil all together to 260 F.

PART 2: Mix the **Nulomoline** with the water, then add the coconut and mix thoroughly.

Add Part 2 to Part 1. Add Part 3. Mix all together until the fondant is dissolved, then add flavors, colors, nuts, fruits, etc. as noted for each flavor below.

BANANA FLAVOR: Add 1 lb **Nulco MeloBan** (powdered banana product) plus yellow color to 25 lb of the batch. Spread the batch out and sprinkle sliced Brazil nuts over the surface.

RASPBERRY FLAVOR: Add raspberry flavor plus pink color, spread the batch out and sprinkle ribbon-cut toasted coconut over the surface.

CHOCOLATE FLAVOR: Add cocoa powder or chocolate liquor to obtain the desired color and flavor. Spread the batch out and top with broken pecans.

VANILLA FLAVOR: Add vanilla flavor and sufficient diced orange, lemon, and citron peel and sprinkle over the surface.

BASIC FONDANT:		
Granulated sugar	80	lb
Corn syrup	20	lb

Invert sugar	10 lb
Water	

Procedure:

Boil granulated sugar, corn syrup, and invert sugar, plus sufficient water to dissolve the sugar, to 242 F (sea level), then cool and cream into fondant.

FRAPPÉ NO. 1:

Corn syrup	50 lb
Invert sugar	50 lb
Albumen	1 lb
Water	2 lb

Procedure:

Cook corn syrup to 245 F (sea level), shut off the heat and add invert sugar and stir until it is melted. Place the batch into a marshmallow beater, start the beater, then gradually add albumen which has been previously dissolved in water and beat the batch until light.

Source: Crompton & Knowles

Creole Rolled Cream Centers

		% by Wt.
Granulated sugar	85 lb	73.82
Molasses (**Grandma's**)	10 lb	8.69
Nulomoline (invert sugar)	10 lb	8.69
Water (to dissolve the sugar)		
Convertit (liquid invertase)	2 oz	0.11
Frappé No. 1 (q.v.)	10 lb	8.69
Nut meats, coconut, etc.		to suit
		Total 100.00

Procedure:

Place the sugar, molasses, **Nulomoline**, and water into a kettle and stir the batch until it boils, then wash down all grains of sugar that adhere to the sides of the kettle and boil the batch to 246–248 F (at sea level).

Pour the batch on a clean cream beater and when it has cooled to approximately 115–120 F, start the beater, add the **Convertit**, and when grain becomes visible, add the Frappé No. 1, nut meats, etc., mixing well.

When the batch sets moderately firm, form it into pieces of the desired size and coat with chocolate.

FRAPPÉ NO. 1:

Corn syrup	50 lb
Nulomoline (invert sugar)	50 lb
Albumen (dissolved in water)	1 lb
Water	2 lb

Procedure:

Boil the corn syrup to 245 F, turn off the heat, add the **Nulomoline** and stir until melted. Promptly place the batch into a marshmallow beater, start the beater, then gradually add the albumen solution and beat the batch until light.

Source: Crompton & Knowles

Crystallized Creams

			% by Wt.
Fondant (q.v.)	100	lb	46.51
BOB:			
Sugar	80	lb	37.21
Corn syrup	10	lb	4.65
Nulomoline (invert sugar)	5	lb	2.33
Water	approx. 20	lb	9.30
			Total 100.00

Procedure:

Prepare fondant.

Boil bob together to 245 F.

Place 100 lb of fondant into a clean cream melting kettle, turn steam on partly. Heat the fondant to approximately 125 F, then add color and flavor; mix well.

Now gradually add boiled bob and mix thoroughly.

Have the hopper of the depositor heated to approximately 160 F.

The temperature of the starch should be approximately 125 F when the batch is cast.

FONDANT:		
Refined cane sugar	80	lb
Corn syrup	15	lb
Nulomoline (invert sugar)	5	lb
Water	approx. 20	lb

Procedure:

Boil altogether to 246 F.

Source: Crompton & Knowles

Crystallized Molasses Creams

No. 1

(Cinnamon Flavor, Red Color)

			% by Wt.
PART 1—BASIC FONDANT (q.v.)	100	lb	50.47
PART 2:			
Molasses (**Grandma's**)	30	lb	15.14
Water	12	lb	6.06
Sugar	50	lb	25.24
Frappé No. 1 (q.v.)	6	lb	3.03
Sucrovert (liquid invertase)	2	oz	0.06
Cinnamon flavor			to suit
			Total 100.00

Procedure:

Prepare fondant.

Place the fondant into a cream melting kettle and heat to approximately 125–130 F.

Meanwhile, boil the sugar, water, and molasses to 248 F. Pour this at once on top of the fondant, mix well, then add the frappé and **Sucrovert** and mix thoroughly. Then work in the flavor.

Cast at once into clean, dry starch.

PART 1—BASIC FONDANT:

Sugar	80	lb
Corn syrup	20	lb
Nulomoline (invert sugar)	10	lb

Procedure:

Cook altogether to 242 F.

FRAPPÉ NO. 1:

Corn syrup	50	lb
Invert sugar	50	lb
Albumen	1	lb
Water	2	lb

Procedure:

Cook corn syrup to 245 F (sea level), shut off the heat and add invert sugar and stir until it is melted. Place the batch into a marshmallow beater, start the beater, then gradually add albumen which has been previously dissolved in water and beat the batch until light.

Source: Crompton & Knowles

No. 2

(Rum Flavor, Daisy Shape)

			% by Wt.
PART 1—BASIC FONDANT (q.v.)	100	lb	50.47
PART 2:			
Molasses (**Grandma's**)	30	lb	15.14
Water	12	lb	6.06
Sugar	50	lb	25.24
Frappé No. 1 (q.v.)	6	lb	3.03
Sucrovert (liquid invertase)	2	oz	0.06
Rum flavor			to suit
			Total 100.00

Procedure:

Prepare basic fondant.

Place the fondant into a cream melting kettle and heat to approximately 125–130 F.

Meanwhile, boil the sugar, water, and molasses to 248 F. Pour this at once on top of the fondant, mix well, then add the frappé and **Sucrovert** and mix thoroughly. Then work in the flavor.

Cast at once into clean, dry starch.

PART 1: BASIC FONDANT:

Sugar	80 lb
Corn syrup	20 lb
Nulomoline (invert sugar)	10 lb

Procedure:

Cook altogether to 242 F.

FRAPPÉ NO. 1:

Corn syrup	50 lb
Invert sugar	50 lb
Albumen	1 lb
Water	2 lb

Procedure:

Cook corn syrup to 245 F (sea level), shut off the heat and add invert sugar and stir until it is melted. Place the batch into a marshmallow beater, start the beater, then gradually add albumen which has been previously dissolved in water and beat the batch until light.

Source: Crompton & Knowles

Extra-Light Cast Cream Centers

		% by Wt.
PART 1—BASIC FONDANT (q.v.)	100 lb	52.58
PART 2:		
Granulated sugar	30 lb	15.77
Corn syrup	20 lb	10.52

Water (to dissolve sugar)			
Frappé No. 1 (q.v.)	40	lb	21.03
Sucrovert (liquid invertase)	3	oz	0.10
Color and flavor			to suit
			Total 100.00

Procedure:

Prepare fondant.

Boil the sugar, corn syrup, and water to 240 F (sea level) and pour this at once into a mixer containing the 100 lb of basic fondant and mix until fondant is melted. Then add the frappé, **Sucrovert**, color, and flavor and continue to stir and heat the batch to 150–155 F. Then cast the batch into large or small starch impressions.

PART 1: BASIC FONDANT

Granulated sugar	80	lb
Corn syrup	20	lb
Nulomoline (invert sugar)	10	lb
Water (to dissolve the sugar)		

Procedure:

Boil altogether to 244 F, cool the syrup to 125 F, then beat the batch into fondant.

FRAPPÉ NO. 1:

Corn syrup	50	lb
Nulomoline (invert sugar)	50	lb
Albumen (dissolved in water)	1	lb
Water	2	lb

Procedure:

Boil the corn syrup to 245 F. Add the **Nulomoline** and stir until it is melted. Promptly transfer the syrup to a marshmallow beater, start the beater, then gradually add the albumen solution and beat the batch until light.

Source: Crompton & Knowles

Extra-Light Rolled Centers

			% by Wt.
Sugar	92	lb	52.22
Nulomoline (invert sugar)	8	lb	4.54
Frappé (q.v.)	20	lb	43.13
Sucrovert (liquid invertase)	3	oz	0.11
			Total 100.00

Procedure:

Cook the sugar and **Nulomoline**, using sufficient water to dissolve the sugar, to 242 F. Start the creaming while the batch is quite hot. As soon as the creaming is started, add the **Sucrovert**, then flavor as desired. Near the end of the creaming, add the frappé and mix well.

This formula produces a short, soft cream center.

NOTE:

For large batches cook the sugar and **Nulomoline** to 244 F.

FRAPPÉ:

Sugar	40	lb
Nulomoline (invert sugar)	10	lb
Nulomoline (invert sugar)	50	lb
Albumen (dissolved in water)	2	lb
Water	4	lb

Procedure:

Cook the sugar and 10 lb of **Nulomoline** to 250 F. Shut off the steam, add the 50 lb of **Nulomoline** (cold) and stir until melted. Place the batch into a marshmallow beater, start the beater, then gradually add the albumen which has been previously dissolved in the water. Beat until light and use as directed above.

Source: Crompton & Knowles

Heavy-Type Cast Cream Centers

			% by Wt.
PART 1—BASIC FONDANT (q.v.)	100	lb	62.43
PART 2:			
Granulated sugar	40	lb	24.97

Corn syrup	10	lb	6.24
Water (to dissolve the sugar)			
Frappé No. 1 (q.v.)	10	lb	6.24
Sucrovert (liquid invertase)	3	oz	0.12
Color and flavor			to suit
			Total 100.00

Procedure:

Prepare fondant.

Boil the sugar, corn syrup, and water to 236–238 F (sea level) and pour this syrup at once into a mixer containing the 100 lb of Basic Fondant and mix until the fondant is melted, then add the frappé, **Sucrovert**, color, and flavor, and mix thoroughly.

Casting temperature of batch 145–150 F.

PART 1—BASIC FONDANT

Granulated sugar	80	lb
Corn syrup	20	lb
Nulomoline (invert sugar)	10	lb
Water (to dissolve sugar)		

Procedure:

Boil altogether to 244 F, cool syrup to 125 F, then beat the batch into fondant.

FRAPPÉ NO. 1:

Corn syrup	50	lb
Nulomoline (invert sugar)	50	lb
Albumen (dissolved in water)	1	lb
Water	2	lb

Procedure:

Boil the corn syrup to 245 F. Add the **Nulomoline** and stir until melted. Promptly transfer the batch to a marshmallow beater, start the beater, then gradually add the albumen solution and beat the batch until light.

Source: Crompton & Knowles

Heavy-Type Cream Centers

(To be Coated in Chocolate Summer Ice-Coat)

			% by Wt.
PART 1—BASIC FONDANT (q.v.)	100	lb	55.52
PART 2:			
Granulated sugar	40	lb	22.21
Corn syrup	20	lb	11.10
Water (to dissolve the sugar)			
PART 3:			
Frappé No. 1 (q.v.)	20	lb	11.10
Sucrovert	2	oz	0.07
			Total 100.00

Procedure:

PART 1: Prepare basic fondant.

PART 2: Boil altogether to 240 F.

PART 3: Add the Frappé No. 1, mix thoroughly, then add **Sucrovert**, color, and flavor, and deposit the batch into starch.

PART 1—BASIC FONDANT

Granulated sugar	80	lb
Corn syrup	20	lb
Nulomoline (invert sugar)	10	lb
Water (to dissolve sugar)		

Procedure:

Heat altogether, stirring the batch occasionally until it boils, then wash down all grains of sugar that adhere to the interior of the kettle and boil the batch rapidly to 244–246 F.

FRAPPÉ NO. 1:

Corn syrup	50	lb
Nulomoline	50	lb
Albumen (dissolved in water)	1	lb
Water	2	lb

Procedure:

Boil the corn syrup to 245 F. Shut off the heat and add the **Nulomoline** and stir until it is melted. Place the batch into a marshmallow beater, start the beater, then gradually add the albumen which has been previously dissolved in water, and beat the batch until light.

Source: Crompton & Knowles

Heavy-Type Commercial Rolled Cream Centers

		% by Wt.
Granulated sugar	85 lb	84.91
Corn syrup	7 lb	6.99
Nulomoline (invert sugar)	8 lb	7.99
Water (to dissolve the sugar)		
Sucrovert (liquid invertase)	1.5 oz	0.09
	Total	100.00

Procedure:

Boil the sugar, corn syrup, **Nulomoline**, and water to 244–246 F (sea level). Pour the batch on a clean beater and when it has cooled to room temperature, start beating. Add **Sucrovert**, color, and flavor as desired; mix well and form the batch into pieces of the desired size and coat with chocolate.

NOTE:

This formula will produce a heavy-type rolled cream center suitable for low price 5 lb bulk packages.

In order to produce a lighter and creamier type of center, add 10 lb of Frappé No. 1, along with the **Sucrovert**, while beating.

Source: Crompton & Knowles

Imitation Hand-Rolled Cream Centers

(Cast into Starch)

		% by Wt.
BASIC FONDANT (q.v.)	100 lb	62.45

BOB:

Sugar	25 lb	15.62
Corn syrup	25 lb	15.62
Frappé (q.v.)	10 lb	6.25
Convertit (invertase)	1.5 oz	0.06
Color and flavor		to suit
		Total 100.00

Procedure:

Prepare fondant.

Cook bob to 236 F.

Pour the bob into 100 lb of the basic fondant and mix thoroughly, then add 10 lb of frappé, and mix well. Now add $1^1/_2$ oz of **Convertit** and mix thoroughly, then add color and flavor as desired. Cast the batch into starch and handle in the usual manner.

NOTE:

For whipped cream centers use 25 lb of frappé instead of the quantity specified above.

BASIC FONDANT:

Coarse granulated sugar	80 lb
Nulomoline (invert sugar)	20 lb
Corn syrup	10 lb
Water (to dissolve the sugar)	

Procedure:

Cook altogether to 240 F, then cool and beat into fondant.

FRAPPÉ:

Nulomoline	50 lb
Nulomoline (cold)	50 lb
Albumen (dissolved in water)	2 lb
Water (cold)	4 lb

Procedure:

Cook 50 lb of **Nulomoline** to 250 F. Shut off the steam and add 50 lb of cold **Nulomoline** and stir until it is melted. Place the batch into a marshmallow beater, start the beater, then gradually add 2 lb of albumen which has been previously dissolved in 4 lb of cold water. Beat until light and use as directed.

Source: Crompton & Knowles

Maple Walnut Cream

			% by Wt.
Brownulated powdered brown sugar	16.75	lb	20.74
Water	8	lb	9.91
Amerfond fondant sugar	50	lb	61.92
Powdered, spray-dried egg albumen	7	oz	0.54
Maple flavor	7	oz	0.54
Invertase	2	oz	0.15
Chopped walnuts	5	lb	6.20
			Total 100.00

Procedure:

Add the water to the powdered brown sugar and heat, with stirring, until dissolved (about 170 F). Transfer brown sugar solution to bowl of planetary mixer. Dry mix the powdered albumen with 17 lb of fondant sugar and, with mixer at low speed, add to the dissolved brown sugar. When blended, turn mixer to high speed and beat for 4 min. Turn mixer to low speed. Add and mix in maple flavor and invertase. Add the balance (33 lb) of fondant sugar and mix thoroughly. Add chopped walnuts and mix until they are evenly distributed. When cool, roll or extrude and coat with chocolate or pastel coating.

Source: Amstar

MeloBan Rolled Cream Centers

			% by Wt.
Sugar	45	lb	73.10
Nulomoline (invert sugar)	5	lb	8.12
Water (to dissolve the sugar)			
Frappé No. 2 (q.v.)	10	lb	16.24
Sucrovert (liquid invertase)	1	oz	0.10
Yellow color			to suit
MeloBan (powdered banana)	1.5	lb	2.44
			Total 100.00

Procedure:

Boil the sugar, **Nulomoline**, and water to 246 F, pour the batch on a clean moist slab, and sprinkle the surface of the batch lightly with water. When it has cooled to approximately 100 F, beat into fondant.

While beating, add **Sucrovert** and sprinkle **MeloBan** into the batch. Add yellow color and when grain develops, work in frappé; mix thoroughly and permit batch to set for 10 min or more, after which it may be formed into pieces of the desired size and shape.

FRAPPÉ NO. 2:

Corn syrup	25	lb
Nulomoline (invert sugar)	25	lb
Albumen (dissolved in the water)	1	lb
Water (cold)	2	lb

Procedure:

Boil the corn syrup to 245 F, add the **Nulomoline** and mix until it is completely dissolved. Turn off the heat, place the batch into a beating machine, start the beater, then gradually add the albumen solution, and beat the batch until it is quite stiff. Remove the frappé from the beater and place into a suitable container for future use.

Source: Crompton & Knowles

Molasses Butter Rolled Cream Centers

			% by Wt.
Granulated sugar	80	lb	65.54
Molasses (**Grandma's**)	20	lb	16.39
Water (to dissolve the sugar)			
Salted dairy butter	10	lb	8.19
Sucrovert (liquid invertase)	1	oz	0.05
Frappé No. 1 (q.v.)	12	lb	9.83
			Total 100.00

Procedure:

Boil the sugar, molasses, and water to 246 F (sea level). Pour the batch on a clean open-type fondant beater and lightly sprinkle the surface of the batch with water and permit it to cool to approximately 90–95 F.

Start the beater, add the **Sucrovert**, and after the batch has mixed for several minutes, add the butter which has been cut into small pieces. After the butter has been worked into the batch, add the frappé and salt, if desired.

Beat the batch until it becomes plastic, then permit it to cool to room temperature and form into pieces of the desired size and shape and coat with chocolate.

FRAPPÉ NO. 1:		
Corn syrup	50	lb
Invert sugar	50	lb
Albumen	1	lb
Water	2	lb

Procedure:

Cook corn syrup to 245 F (sea level), shut off the heat and add invert sugar and stir until it is melted. Place the batch into a marshmallow beater, start the beater, then gradually add albumen which has been previously dissolved in water and beat the batch until light.

Source: Crompton & Knowles

Molasses Cast Cream Centers

(For Bulk or Bar Goods, Vanilla Flavor)

			% by Wt.
PART 1—BASIC FONDANT (q.v.)	100	lb	57.43
PART 2:			
Molasses (**Grandma's**)	20	lb	11.49
Water	4	lb	2.30
Corn syrup	10	lb	5.74
Sugar	10	lb	5.74
Frappé No. 3 (q.v.)	30	lb	17.23
Sucrovert (liquid invertase)	2	oz	0.07
Vanilla flavor			to suit
			Total 100.00

Procedure:

PART 1: Prepare basic fondant.

PART 2: Place the fondant into a cream melting kettle and heat the batch to approximately 120 F.

Meanwhile, boil the molasses, water, and sugar to 234 F, then pour this at once on top of the fondant and mix thoroughly. Now add the frappé and **Sucrovert**, mix well, then work in the flavor.

Casting temperature approximately 140–145 F.

PART 1—BASIC FONDANT:	
Sugar	80 lb
Corn syrup	20 lb
Nulomoline (invert sugar)	10 lb

Procedure:

Cook together to 242 F.

FRAPPÉ NO. 3:	
Corn syrup	50 lb
Nulomoline	50 lb
Gelatin	1.5 lb
Water	3 lb
Albumen	1.5 lb
Water	3 lb

Procedure:

Boil corn syrup to 245 F. Shut off the steam, add **Nulomoline**, and stir until melted. Place the batch into a beater, start the beater, then add gelatin which has been soaked in 3 lb of water, and beat until dissolved. Then gradually add albumen which has been previously dissolved in 3 lb of water and beat the batch until it is light.

Source: Crompton & Knowles

Molasses Cream Squares

		% by Wt.
Granulated sugar	50 lb	38.46
Molasses (**Grandma's**)	15 lb	11.54

Corn syrup	25	lb	19.23
Medium shred coconut	10	lb	7.69
Coconut butter (94 F m.p.)	5	lb	3.85
Basic fondant (q.v.)	25	lb	19.23
		Total	100.00

Procedure:

Cook the sugar, molasses, and corn syrup to 244 F. Shut off the steam and add the coconut. Mix this through the batch thoroughly at the same time adding the coconut butter. After these are mixed, add the basic fondant, then pour and handle as usual.

BASIC FONDANT:

Sugar	80	lb
Corn syrup	20	lb
Nulomoline (invert sugar)	10	lb

Procedure:

Cook together to 242 F.

Source: Crompton & Knowles

Molasses Cream Thin Mints with Chocolate Coating

No. 1

(Short Molasses Cream Mints)

			% by Wt.
Fondant (q.v.)	100	lb	95.24
Frappé No. 1 (q.v.)	5	lb	4.76
		Total	100.00

Procedure:

Melt 100 lb of the fondant, heating to 160 F. Add 5 lb of Frappé No. 1; mix thoroughly. Cast onto waxed paper to form wafers or into rubber molds. When cold, turn upside down, dry the bottoms and coat with chocolate.

FONDANT:

Sugar	75	lb
Corn syrup	25	lb

Procedure:

Cook to 245 F.

FRAPPÉ NO. 1:

Corn syrup	50	lb
Invert sugar	50	lb
Albumen	1	lb
Water	2	lb

Procedure:

Cook corn syrup to 245 F (sea level), shut off the heat and add invert sugar and stir until it is melted. Place the batch into a marshmallow beater, start the beater, then gradually add albumen which has been previously dissolved in water and beat the batch until light.

No. 2

(Moderately Soft Molasses)

Fondant (q.v.)	100	lb	90.81
Frappé No. 1 (q.v.)	10	lb	9.08
Convertit (invertase)	2	oz	0.11
			Total 100.00

Procedure:

Melt 100 lb of the fondant to 155 F. Add 10 lb of Frappé No. 1 and 2 oz of **Convertit** and mix thoroughly. Cast into starch impressions and when cold, remove from starch; coat in chocolate.

FONDANT:

Granulated sugar	100	lb
Molasses (**Grandma's**)	20	lb
Corn syrup	5	lb
Water (to dissolve the sugar)		

Procedure:

Boil sugar, molasses, corn syrup, and water to 242 F (sea level). Pour batch out on a clean, cool cream beater and lightly sprinkle surface of the hot batch with water. When the batch cools to approximately 90 F, beat into fondant.

FRAPPÉ NO. 1:

Corn syrup	50	lb
Nulomoline (invert sugar)	50	lb
Albumen (dissolved in the water)	1	lb
Water	2	lb

Procedure:

Boil the corn syrup to 245 F. Turn off the heat and add **Nulomoline**; stir until melted. Place the batch into a marshmallow beater, start the beater, then gradually add the albumen solution and beat the batch until light.

Source: Crompton & Knowles

Molasses Easter Eggs

			% by Wt.
Granulated sugar	35	lb	58.15
Molasses (**Grandma's**)	15	lb	24.92
Sucrovert	1.5	oz	0.16
Fine table salt	1.5	oz	0.16
Frappé (q.v.)	5	lb	8.31
Coconut preparation (q.v.)	5	lb	8.30
		Total	100.00

Procedure:

Cook the sugar and molasses together, adding just enough water to dissolve the sugar. The cooking temperature will have to be varied to suit the size of egg wanted. A temperature of 246 F is suggested. Pour on an open beater, such as the Ball, and when almost cold, start creaming. Add $1^1/_2$ oz of **Sucrovert**. When this has been mixed through the batch, add $1^1/_2$ oz of fine table salt. Just before the creaming is finished add 5 lb of frappé and 5 lb of short cut coconut. When the creaming is finished, form the centers, and when cool, coat them with chocolate.

FRAPPÉ:

Corn syrup	25	lb
Nulomoline (cold)	25	lb
Albumen (dissolved in the water)	0.75	lb
Water (cold)	1.5	lb

Procedure:

Cook the corn syrup to 248 F; then shut off the steam. Add the cold **Nulomoline** and stir until melted. Then beat the batch with the albumen previously dissolved in cold water.

COCONUT PREPARATION:

Coconut	4	pbw
Nulomoline	1	pbw

Procedure:

Mix 4 pbw of coconut with 1 pbw of melted **Nulomoline** and stir together without heating. Make this mixture the day before it is to be used.

Source: Crompton & Knowles

Molasses Mint Cast Cream Centers

No. 1

			% by Wt.
Basic Fondant (q.v.)	100	lb	57.10
Molasses (**Grandma's**)	35	lb	19.99
Granulated sugar	20	lb	11.41
Water	5	lb	2.86
Frappé No. 1 (q.v.)	15	lb	8.57
Sucrovert (liquid invertase)	2	oz	0.07
Mint flavor			to suit
			Total 100.00

Procedure:

Boil the molasses, sugar, and water to 242 F and pour this at once on top of Basic Fondant, mixing thoroughly, then add the mint flavor, mixing well. Casting temperature is approximately 145–150 F.

BASIC FONDANT:

Granulated sugar	80	lb
Corn syrup	20	lb
Nulomoline	10	lb
Water (to dissolve sugar)		

Procedure:

Boil the sugar, corn syrup, **Nulomoline**, plus sufficient water to dissolve the sugar to 244 F, cool to 125 F, then beat into fondant.

FRAPPÉ NO. 1:

Corn syrup	50	lb
Nulomoline (invert sugar)	50	lb
Albumen (dissolved in the water)	1	lb
Water	2	lb

Procedure:

Boil the corn syrup to 245 F. Add the **Nulomoline** and mix until the **Nulomoline** is dissolved. Promptly transfer this syrup to a marshmallow beater, start the beater, then gradually add 1 lb of albumen which has previously been dissolved in 2 lb of water and beat the batch until light.

Source: Crompton & Knowles

No. 2

(For Bulk, Package and Bar Goods)

			% by Wt.
PART 1—BASIC FONDANT (q.v.)	100	lb	58.78
PART 2—BOB:			
Molasses (**Grandma's**)	20	lb	11.76
Granulated sugar	20	lb	11.76
Water	5	lb	2.94

Corn syrup	10	lb	5.88
Frappé No. 1 (q.v.)	15	lb	8.81
Sucrovert (liquid invertase)	2	oz	0.07
Mint oil			to suit
			Total 100.00

Procedure:

PART 1: Prepare fondant.

PART 2—BOB: Boil the molasses, sugar, water, and corn syrup to 236–238 F.

Pour this syrup into 100 lb of Basic Fondant (Part 1) mixing well, then add the Frappé, **Sucrovert,** and mint oil, mix thoroughly, then cast into starch impressions.

When making large size centers, it is suggested that the bob be boiled to 242 F.

Casting temperature for bulk and package items is 145–150 F.

Casting temperature for larger items is 155–165 F.

PART 1—BASIC FONDANT:

Granulated sugar	80	lb
Corn syrup	20	lb
Nulomoline (invert sugar)	10	lb
Water (sufficient to dissolve the sugar)		

Procedure:

Boil altogether to 242–244 F, cool and beat into fondant.

FRAPPÉ NO. 1:

Corn syrup	50	lb
Invert sugar	50	lb
Albumen	1	lb
Water	2	lb

Procedure:

Cook corn syrup to 245 F (sea level), shut off the heat and add invert sugar and stir until it is melted. Place the batch into a marshmallow beater, start the beater, then gradually add albumen which has been previously dissolved in water and beat the batch until light.

Source: Crompton & Knowles

Molasses Whipped Cream Centers

			% by Wt.
Granulated sugar	70	lb	53.79
Water	15	lb	11.53
Molasses (**Grandma's**)	30	lb	23.05
Sucrovert (liquid invertase)	2	oz	0.10
Frappé No. 2 (q.v.)	15	lb	11.53
			Total 100.00

Procedure:

Place the molasses, sugar, and water into a kettle. Stir the batch occasionally until it reaches the boiling point, then wash down sides of kettle to remove all grains of sugar. Place a thermometer into the batch and boil to 244 F. Pour on a properly cooled beater; and when the batch has cooled to approximately 95 F, start the beater, then add the **Sucrovert**. When a slight grain becomes visible, add the frappé and permit the batch to beat until it has completely grained.

Remove the batch from the beater and cool it before rolling into pieces of the desired shape. Then coat the centers in chocolate.

FRAPPÉ NO. 2:

Corn syrup	25	lb
Nulomoline (invert sugar)	25	lb
Albumen (dissolved in the water)	1	lb
Water (cold)	2	lb

Procedure:

Boil the corn syrup to 245 F, add the **Nulomoline** and mix until it is completely dissolved. Turn off the heat, place the batch into a beating machine, start the beater, then gradually add the albumen solution, and beat the batch until it is quite stiff. Remove the frappé from the beater and place into a suitable container for future use.

Source: Crompton & Knowles

Nulco MeloBan Rolled Creams

			% by Wt.
Sugar	85	lb	71.35
Nulomoline (invert sugar)	15	lb	12.59
Water (to dissolve the sugar)			
Nulco MeloBan (powdered ripe banana)	4	lb	3.36
Citric or tartaric acid solution	1	oz	0.05
Banana color			to suit
Convertit (invertase)	1	oz	0.05
Special Frappé (q.v.)	15	lb	12.60
			Total 100.00

Procedure:

Boil sugar, **Nulomoline**, and water to 244–248 F (sea level); cool batch to 100–110 F and while beating it into fondant, sift in **Nulco MeloBan**, color, **Convertit**, and acid solution. When batch develops grain, add frappé, mix well, and form into piece of the desired size.

SPECIAL FRAPPÉ:

Corn syrup	25	lb
Sugar	25	lb
Water (to dissolve sugar)		
Nulomoline	25	lb
Corn syrup	25	lb
Albumen (dissolved in the water)	2	lb
Water	4	lb

Procedure:

Cook the corn syrup, sugar, and water to 245 F. Add the **Nulomoline** and remaining 25 lb of corn syrup and mix until the **Nulomoline** is melted. Place batch into a marshmallow beater, start the beater; then gradually add the albumen solution and beat until light.

Source: Crompton & Knowles

Nut Butter Creams

			% by Wt.
Granulated sugar	20	lb	61.42
Salt	2	Tbl	0.19
Corn syrup	2	lb	6.14
Butter	1	lb	3.07
Light cream	1	gal	26.10
Chopped walnuts	1	lb	3.08
		Total	100.00

Procedure:

Cook sugar, corn syrup, cream, and salt to 241 F, stirring constantly. Pour on clean slab, sprinkled lightly with water. When almost cold, work into fondant with cream scraper, adding gradually the butter and chopped nuts. Then roll into small balls and dip in chocolate.

Great care must be exercised when cooking to keep sides of your kettle washed down, so that no syrup may grain on candy.

Source: Savage Brothers Co.

Pacific Coast Hand-Rolled Creams

			% by Wt.
Coarse granulated sugar	90	lb	66.13
Nulomoline (invert sugar)	10	lb	7.35
Water	24	lb	17.63
Special Frappé (q.v.)	12	lb	8.82
Sucrovert (liquid invertase)	1.5	oz	0.07
		Total	100.00

Procedure:

Boil sugar, **Nulomoline**, and water to 241–243 F, then pour on a clean cream beater on which has been spread 1 lb of **Nulomoline.**

Sprinkle surface of the batch with 3 oz of water and cool to 80 F. Add **Sucrovert**, start the machine and when batch shows signs of graining, add half of frappé and flavor. When the batch begins to work free from the mixers, add the balance of the frappé, and mix well.

Stop the machine and immediately roll the batch into strips for hand rolling. If a center forming machine is used, the fondant should be cooled to approximately 70 F before being placed in the machine. Coat centers with chocolate as soon as convenient.

SPECIAL FRAPPÉ:

Corn syrup	50	lb
Nulomoline (invert sugar)	50	lb
Albumen (dissolved in 5 lb water)	2.5	lb
Water	5	lb
Tartaric acid (mixed with 2 oz water)	1	oz
Water	2	oz

Procedure:

Boil the corn syrup to 245 F (sea level); turn off heat, add the **Nulomoline** (cold) and mix well. Gradually add the albumen solution; then acid solution and beat the batch until light and use as directed.

Source: Crompton & Knowles

Pineapple Juice Cream

			% by Wt.
Amerfond fondant sugar	50	lb	84.43
Powdered, spray-dried egg albumen	7	oz	0.74
Frozen concentrated pineapple juice* (thawed)	6 lb 4	oz	10.55
Color*			to suit
Citric acid	2.5	oz	0.26
Water	2 lb 5	oz	3.91
Invertase	1	oz	0.11
			Total 100.00

Procedure:

Place 25 lb of fondant sugar in the bowl of planetary mixer and, at low speed, blend in the powdered albumen, then slowly add the thawed pineapple juice and color.

* Other frozen concentrated juices and appropriate colors may be substituted.

Dissolve the citric acid in the water and mix into the batch, then turn to high speed and beat for 4 min.

Turn mixer to low speed, add invertase and balance (25 lb) of fondant sugar and mix until smooth.

Roll or extrude and coat with chocolate or pastel coating.

Source: Amstar

Plastic Cast Cream Centers

			% by Wt.
PART 1—BASIC FONDANT (q.v.)	100	lb	60.56
PART 2:			
Granulated sugar	30	lb	18.17
Corn syrup	20	lb	12.11
Water (to dissolve sugar)			
Frappé No. 1 (q.v.)	15	lb	9.08
Sucrovert (liquid invertase)	2	oz	0.08
Color and flavor			to suit
			Total 100.00

Procedure:

PART 1: Prepare basic fondant.

PART 2: Boil the sugar, corn syrup, and water to 238 F (sea level). Pour this at once into a mixer containing the 100 lb of Basic Fondant and mix until fondant is melted. Now add frappé, **Sucrovert**, color, and flavor; continue to stir and heat batch to 150–155 F, then cast the batch into large or small starch impressions.

BASIC FONDANT:		
Granulated sugar	80	lb
Corn syrup	20	lb
Nulomoline (invert sugar)	10	lb
Water (to dissolve the sugar)		

Procedure:

Boil altogether to 244 F, cool syrup to 125–130 F then beat the batch into fondant.

FRAPPÉ NO. 1:

Corn syrup	50	lb
Nulomoline (invert sugar)	50	lb
Albumen (dissolved in the water)	1	lb
Water	2	lb

Procedure:

Boil the corn syrup to 245 F. Add the **Nulomoline** and mix until the **Nulomoline** is dissolved. Promptly transfer this syrup to a marshmallow beater, start the beater, then gradually add 1 lb of albumen which has previously been dissolved in 2 lb of water and beat the batch until light.

Source: Crompton & Knowles

Plastic Chocolate Centers

			% by Wt.
Liquor chocolate (unsweetened)	50	lb	36.99
Nulomoline (invert sugar)	60	lb	44.38
Sucrovert (liquid invertase)	1	oz	0.05
Frappé No. 1 (q.v.)	25	lb	18.48
Mint oil	1	oz	0.05
Powdered salt	1	oz	0.05
			Total 100.00

Procedure:

Melt the liquor chocolate and temper it to approximately 90 F.

Meanwhile, heat the **Nulomoline** until dissolved and permit it to cool to approximately 75–90 F, then add the cooled **Nulomoline** and **Sucrovert** to the tempered chocolate and mix until smooth. Add the mint oil and salt, then the Frappé No. 1 and mix thoroughly.

Spread the batch on waxed paper to set, after which it may be extruded through a rolled center machine, or cut into pieces of the desired size or rolled by hand to form small or large centers which may later be coated with chocolate and rolled in nuts.

Sucrovert is added to check the development of mold.

FRAPPÉ NO. 1:

Corn syrup	50	lb
Nulomoline (invert sugar)	50	lb
Albumen (dissolved in the water)	1	lb
Water	2	lb

Procedure:

Boil the corn syrup to 245 F (sea level). Turn off the heat and add the **Nulomoline** and stir until it is melted. Promptly transfer this syrup to a marshmallow beater, start the beater, then gradually add 1 lb of albumen which has previously been dissolved in 2 lb of water and beat the batch until light.

Source: Crompton & Knowles

Plastic Rolled Cream Centers

			% by Wt.
Granulated sugar	90	lb	74.93
Nulomoline (invert sugar)	10	lb	8.32
Water (to dissolve the sugar)			
Frappé No. 2 (q.v.)	20	lb	16.65
Sucrovert (liquid invertase)	2	oz	0.10
Color and flavor			to suit
			Total 100.00

Procedure:

Bring the sugar, **Nulomoline**, and water to the boiling point. Wash down all grains of sugar that adhere to the kettle and boil the batch rapidly to 250 F (sea level).

Pour the batch at once onto a clean cream beater and when it has cooled to approximately 105–110 F start the beater. Add the **Sucrovert** and about 10 lb of the frappé; after the batch has beaten for about 5 min, add color and flavor and the balance of the frappé and mix well.

When the batch has set moderately firm, permit it to cool to room temperature; then form into pieces of the desired size and shape and coat with light or dark chocolate.

FRAPPÉ NO. 2:

Corn syrup	50	lb
Nulomoline (invert sugar)	50	lb
Albumen (dissolved in the water)	2	lb
Water	4	lb

Procedure:

Boil the corn syrup to 245 F. Add the **Nulomoline** and stir until it is melted. Promptly transfer this syrup to a marshmallow beater, start the beater, then gradually add the albumen solution and beat the batch until light.

Source: Crompton & Knowles

Quick-Set Cream Centers

			% by Wt.
PART 1—FONDANT (q.v.)	100	lb	76.83
PART 2—BOB:			
Sugar	19	lb	14.60
Nulomoline (invert sugar)	1	lb	0.77
PART 3:			
Frappé (q.v.)	10	lb	7.68
Convertit (invertase)	2.5	oz	0.12
		Total	100.00

Procedure:

PART 1: Prepare fondant.

PART 2—BOB: Cook sugar and **Nulomoline**, together with sufficient water to dissolve sugar, to 244 F.

Stir into the bob 100 lb of fondant, add 10 lb of Frappé and heat batch to 150 F. Now add the **Convertit**, taking care that it does not come in contact with the sides of kettle and when **Convertit** has been well mixed into batch, add flavor and cast the batch into starch.

NOTE:

This process must be worked continuously from the time of cooking until the centers are cast. As soon as fondant is cooled it should be immediately cooked and beaten; when it comes from the cream beater it should be placed into a remelt kettle and the bob added. If this procedure is not followed the fondant will become so soft that it will be impossible to get it to set again. The batch should not be held in the remelt kettle any longer than is necessary to mix and bring to the desired degree of heat. When ready it should be put into the depositor and cast.

FONDANT:

Sugar	85 lb
Nulomoline (invert sugar)	15 lb
Water	22 lb

Procedure:

Cook all of the ingredients together to 238 F, cool to approximately 115 F, then beat into fondant.

FRAPPÉ:

Sugar	40 lb
Nulomoline (invert sugar)	60 lb
Albumen (soaked in the water)	1.5 lb
Water	3 lb

Procedure:

Place sugar and 10 lb of **Nulomoline**, together with sufficient water to dissolve sugar, into a kettle and cook to 244 F. Shut off the heat, add the remaining 50 lb of **Nulomoline** and stir until melted. Place batch into a beater of the Hobart type, gradually add albumen solution, and beat until 1 qt of the frappé weighs 1 lb.

Source: Crompton & Knowles

Rolled Cream Centers

No. 1

		% by Wt.
Coarse granulated sugar	90 lb	81.73
Water (to dissolve the sugar)		

Nulomoline (invert sugar)	10	lb	9.08
Frappé No. 1 (q.v.)	10	lb	9.08
Sucrovert (liquid invertase)	2	oz	0.11
			Total 100.00

Procedure:

Cook the sugar, **Nulomoline,** and water to 242–246 F which makes a fondant that will produce a soft center. After cooking, cool to a temperature of 110 F, then cream. As soon as the creaming is started, add the **Sucrovert,** then flavor and color. Near the end of the creaming, add the frappé.

FRAPPÉ NO. 1:

Corn syrup	50	lb
Nulomoline (invert sugar)	50	lb
Albumen (dissolved in the water)	1	lb
Water	2	lb

Procedure:

Boil the corn syrup to 245 F. Shut off steam and add the **Nulomoline** (cold) and stir until the **Nulomoline** is melted. Place the batch into a marshmallow beater, then gradually add the albumen solution and beat the batch until light.

Source: Crompton & Knowles

No. 2

(Containing Cream of 20% Butter Fat)

			% by Wt.
Granulated sugar	90	lb	65.38
Nulomoline (invert sugar)	10	lb	7.27
Cream containing 20% butter fat	25	lb	18.17
Water	2	lb	1.45
Frappé No. 2 (q.v.)	10	lb	7.27
Salt	5	oz	0.23
Vanilla flavor	3	oz	0.14
Sucrovert (liquid invertase)	2	oz	0.09
			Total 100.00

Procedure:

Boil sugar, **Nulomoline**, water, and cream together to approximately 244 F. Cool and cream to 110 F. Start the machine and add the **Sucrovert**. When the **Sucrovert** has been thoroughly mixed through the batch, add flavor and salt. When the batch begins to show evidence of graining, add the frappé. Do not beat the batch any more than is necessary.

FRAPPÉ NO. 2:

Corn syrup	50	lb
Nulomoline (invert sugar)	50	lb
Albumen (dissolved in the water)	2	lb
Water (cold)	4	lb

Procedure:

Boil the corn syrup to 245 F. Add the **Nulomoline** at once, mix well, then turn off the heat. Place the batch into a marshmallow beater, start the beater, then gradually add the albumen solution and beat until quite light.

Source: Crompton & Knowles

Rolled Fruit-Nut Bars and Eggs

			% by Wt.
Granulated sugar	40	lb	51.10
Nulomoline (invert sugar)	5	lb	6.39
Corn syrup	5	lb	6.39
Water	10	lb	12.77
Frappé	3	lb	3.83
Sucrovert (liquid invertase)	0.5	oz	0.04
Salt	0.25	lb	0.32
Assorted preserved fruits and nuts	15	lb	19.16
			Total 100.00

Procedure:

Place the sugar, corn syrup, **Nulomoline**, and water into a kettle. Apply heat, bring the batch to the boiling point. Wash down all crystals of sugar that adhere to the kettle, and cool the batch as rapidly as possible to 250–254 F.

Pour the batch onto an open-type fondant beater. Lightly sprinkle the surface of the syrup with water, and permit the batch to cool to approximately 125 F.

Start the beater, add the **Sucrovert**, salt, and flavor as desired. Continue to beat until the batch becomes opaque, then add the frappé and after the frappé has been well distributed add the fruit and nuts. Continue to beat until the batch becomes plastic.

Form the batch into egg shapes or bars and as soon as the temperature of the formed pieces is under 75–80 F coat with chocolate.

Source: Crompton & Knowles

Rolled or Cast Cream Easter Eggs for Slices

		% by Wt.
PART 1—BASIC FONDANT (q.v.)	100 lb	42.18
PART 2—BOB:		
Granulated sugar	50 lb	21.09
Corn syrup	25 lb	10.54
Water	12 lb	5.06
PART 3:		
Frappé No. 1 (q.v.)	20 lb	8.44
Fruits, nuts	30 lb	12.65
Sucrovert	1.5 oz	0.04
		Total 100.00

Procedure:

PART 1: Prepare fondant.

PART 2—BOB: Boil altogether to 246–250 F.

Pour cooked bob syrup into the fondant, mix well, then add 20 lb of Frappé No. 1, flavor, and mix thoroughly; then add 20–30 lb of assorted preserved fruits, nuts, etc., which have first been boiled in **Nulomoline** for several minutes, then drained. (Boil the fruits and nuts in **Nulomoline** before starting the batch so that they will be cold at the time they are added.)

Check on the temperature of the batch which should be 160–165 F, then add $1^{1}/_{2}$ oz of **Sucrovert** and mix thoroughly.

The batch is now ready to be cast into starch or it may be spread out to cool and later rolled or formed into bars or egg shapes.

BASIC FONDANT:

Granulated sugar	80	lb
Corn syrup	20	lb
Nulomoline (invert sugar)	10	lb
Water (sufficient to dissolve sugar)		

Procedure:

Boil altogether to 242–246 F, cool, then beat into fondant in the usual manner.

FRAPPÉ NO. 1:

Corn syrup	50	lb
Nulomoline (invert sugar)	50	lb
Albumen (dissolved in the water)	1	lb
Water	2	lb

Procedure:

Boil the corn syrup to 245 F, shut off the heat, add the **Nulomoline** and stir until it is melted. Place the batch into a beater, start the beater, then gradually add the albumen solution and beat the batch until light.

Source: Crompton & Knowles

Rolled Peanut Butter Cream Centers

			% by Wt.
Granulated sugar	80	lb	57.09
Nulomoline (invert sugar)	10	lb	7.14
Water	20	lb	14.27
Peanut butter	15	lb	10.70
Molasses (**Grandma's**)	5	lb	3.57
Frappé No. 1 (q.v.)	10	lb	7.14
Sucrovert (liquid invertase)	2	oz	0.09
Salt and flavor			to suit
			Total 100.00

Procedure:

Place the sugar, **Nulomoline**, and water into a kettle. Apply heat, bring to a boil, wash down the sides of the kettle of all sugar crystals. Cook to 244–246 F. Pour on cool, clean cream beater. Allow to cool to 100–110 F. Start the beater. When the batch shows evidence of graining, add the peanut butter, molasses, frappé, **Sucrovert**, salt, and flavor. Beat until the batch becomes firm.

When cool, form the centers by machine or hand. Coat with chocolate, icing, or fondant.

FRAPPÉ NO. 1:

Corn syrup	50 lb
Invert sugar	50 lb
Albumen	1 lb
Water	2 lb

Procedure:

Cook corn syrup to 245 F (sea level), shut off the heat and add invert sugar and stir until it is melted. Place the batch into a marshmallow beater, start the beater, then gradually add albumen which has been previously dissolved in water and beat the batch until light.

Source: Crompton & Knowles

Rum 'n Almond Patties

		% by Wt.
White sugar	50 lb	46.71
Brown sugar	5 lb	4.67
Corn syrup	6 lb	5.61
Manufacturing cream (20% butterfat)	1.5 gal	11.91
Water	0.5 gal	3.90
Salt	2 oz	0.12
Frappé	4 lb	3.74
Buttered-diced roasted almonds	25 lb	23.34
		Total 100.00

Procedure:

Cook sugar, corn syrup, cream, and water to 242 F. Add salt at any time after the boil. Pour on cream beater and allow to cool. Add flavor while the batch is cooling. Add frappé and almonds just before the batch comes up. Form in flat round patties and dip in milk and dark chocolate.

Source: California Almond Growers Exchange

Sliced Cream Eggs and Bars

		% by Wt.
Granulated sugar	8 lb	72.52
Nulomoline (invert sugar)	2 lb	18.13
Water (to dissolve the sugar)		
Frappé No. 1 (q.v.)	1 lb	9.07
Sucrovert (liquid invertase)	0.5 oz	0.28
Preserved fruits and nuts		to suit
Color and flavor		to suit
		Total 100.00

Procedure:

Place sugar, **Nulomoline**, and water into a kettle; bring the batch to boiling point. Wash down all grains of sugar that adhere to the interior of the kettle and boil the batch rapidly to 250 F (sea level).

Pour batch on a damp slab or cream beater to partly cool; when batch has cooled to 115–120 F beat into fondant. While beating, add frappé, mix well; then add **Sucrovert**, color, and flavor, preserved fruits and nuts, as desired.

When batch is cool, form it into bars or eggs of suitable size. Allow to set until the surface dries, then coat with chocolate.

FRAPPÉ NO. 1:

Corn syrup	10 lb
Nulomoline (invert sugar)	10 lb
Albumen (dissolved in the water)	4 oz
Water	8 oz

Procedure:

Boil the corn syrup to approximately 245 F (sea level); turn off the heat, add the **Nulomoline** and stir until it is melted. Place batch into a marshmallow beater, start the beater, then gradually add the albumen solution and beat the batch until light.

Source: Crompton & Knowles

Special Rolled Coconut Creams

(For Bars)

			% by Wt.
FONDANT:			
Refined cane sugar	44	lb	56.37
Nulomoline (invert sugar)	6	lb	7.69
Water	15	lb	19.20
Coconut paste (q.v.)	8	lb	10.25
Convertit (invertase)	1	oz	0.08
Frappé	5	lb	6.41
Vanilla			to suit
			Total 100.00

Procedure:

FONDANT: Place the water, sugar, and **Nulomoline** into a clean kettle. Stir the batch occasionally until it reaches the boiling point. Wash down the sides of the kettle to remove all grains of sugar. Place a wooden cover on top of the kettle and permit the batch to boil rapidly for a few minutes, then partly remove the cover and insert a tested thermometer. Boil the batch as rapidly as possible to 250 F. Turn off the heat.

Meanwhile spread the 8 lb of the coconut paste over the ball beater. Pour the batch on top of the coconut paste and sprinkle the batch with not more than 2 oz of water. Allow the batch to cool to approximately 100 F.

Now start the machine in motion, add 1 oz **Convertit** and permit the batch to beat until a slight grain becomes visible, then add 5 lb of frappé and sufficient vanilla, scraping sides and mixing arms of beater, and when the batch turns up firm stop the beater. Place the batch on cooling slabs so that it will cool to room temperature. Do not attempt to run the batch through the machine until it is cool.

COCONUT PASTE:

			% by Wt.
Long shred coconut	12	lb	37.50
Nulomoline (invert sugar)	18	lb	56.25
Water	2	lb	6.25
			Total 100.00

Procedure:

Place the water and **Nulomoline** into a kettle and beat until the **Nulomoline** is completely dissolved (do not permit the batch to boil). Add the coconut and mix thoroughly. Heat the batch slightly for a minute or two and set aside for at least 2 h.

It is suggested that a stock of this preparation be kept on hand as it is the function of **Nulomoline** to soften the coconut, thus making it tender and preventing it from absorbing moisture from the fondant.

Source: Crompton & Knowles

Chapter 2

AERATED CANDIES, MISCELLANEOUS

Creamed Mints

No. 1

(Boiled in Simplex Vacuum Cooker)

			% by Wt.
Granulated sugar	85	lb	67.46
Nulomoline (invert sugar)	15	lb	11.90
Water	24	lb	19.05
Oil of peppermint			to suit
Pulverized sugar (for seeding purposes)	2	lb	1.59
Blue color			trace
			Total 100.00

Procedure:

Place into a clean kettle the water, **Nulomoline**, and granulated sugar and mix the batch well before turning on the heat. Continue to stir the batch until it reaches the boiling point, then carefully wash down the sides of the kettle to remove all grains of sugar and boil the batch to 248 or 250 F (sea level). Turn the heat off, then place the hood on the kettle and start the vacuum and after the vacuum gauge reaches 15 in. allow the vacuum to pull for 4 min. Release the vacuum and spread the batch out to cool. When the batch is partially cooled work into it the pulverized sugar, mint oil, and blue color. Place the batch on the pulling machine and pull approximately 10 min. Dust the spinning table with powdered sugar and run the batch through rollers. Spread the mints in single layers onto starch trays, which have been well sprinkled with pulverized sugar. Do not disturb the mints until they have grained.

Sieve off the surplus sugar and pack the mints in tins lined with waxed paper, using an atomizer to spray a little water into the half-filled tins of mints. Finish filling the containers with the mints and tightly seal the tins.

(For each 550 ft above sea level reduce the boiling temperature by 1 F.)

Source: Crompton & Knowles

No. 2

(Open Fire)

			% by Wt.
Granulated sugar	85	lb	67.46
Nulomoline (invert sugar)	15	lb	11.90
Cold water	24	lb	9.05
Oil of peppermint			to suit
Pulverized sugar (for seeding purposes)	2	lb	1.59
Blue color			trace
			Total 100.00

Procedure:

Place the water, **Nulomoline**, and granulated sugar into a clean kettle. Set the kettle on a large forced draught gas furnace and stir well before lighting the gas. Turn on the gas about half. Continue to stir the batch well until it reaches the boiling point.

Carefully wash down all grains of sugar that adhere to the sides of the kettle. Place a wooden cover over the batch so as to steam down the sides of the kettle. Allow the batch to steam for 4 or 5 min, then remove the cover from the kettle and insert a tested thermometer. Boil the batch as rapidly as possible to 275 F. Then pour it on a large cooler which should be absolutely clean and quite cold. Allow the batch to remain undisturbed for 4 or 5 min, then carefully fold in the cooled edges and as the batch cools continue to fold together, but care must be taken so as not to work the batch any more than is absolutely necessary as this might cause premature graining.

Remove about 10 lb of the batch and work into it the mint oil, the pulverized sugar, and the blue color. When the batch is cool enough to place on the pulling machine, add the portion containing the pulverized sugar. Pull the batch from 8–10 min on a standard pulling machine.

After pulling, run through the cutting rollers and let the mints remain undisturbed until they have grained. Sieve all the webs using a coarse sieve for this purpose, then pack the mints into tins lined with waxed paper. Fill the tins about half full, spray a little moisture into the tins using an atomizer for this purpose, then finish filling.

(For each 550 ft above sea level, reduce the boiling temperature by 1 F.)

Source: Crompton & Knowles

Grained (Creamed) Mints

		% by Wt.
High-grade refined cane sugar	30 lb	74.08
Water	8 lb	19.75
Nulomoline (invert sugar)	2.5 lb	6.17
		Total 100.00

Procedure:

Place the sugar, water, and **Nulomoline** into a clean kettle and mix well. Apply heat (forced draught gas furnace), stir the batch occasionally until it boils, then carefully wash down all grains of sugar that cling to the kettle. Place a metal or wooden cover (steamer) over the kettle and allow the batch to cook for 3 or 4 min, then remove the steamer. Place a tested thermometer into the batch and boil the batch as rapidly as possible to 278–280 F.

Meanwhile, a clean, cool metal slab should be oiled and the cooked batch poured onto the cooling slab. Do not disturb the batch for 3 or 4 min to permit it to partially cool, then carefully fold in about 4 in. of the edges, then take a long knife passing it under the batch to release it from the slab. Allow the batch to cool for another 2 or 3 min, then carefully fold it together but do not work or knead it. As the batch rapidly cools, fold it together from time to time and when it is fairly firm, place it on the pulling hook, pulling for not longer than 4 min, using a power puller. Pull for 7 or 8 min if pulling by hand. While pulling add approximately $^3/_4$–1 oz of peppermint oil, plus $1^1/_2$ oz of glycerin.

After the batch has been properly pulled, knead it well to force out some of the air, then either spin it out by hand or pass it through rollers or a standard grained mint cutter.

The cut mints are then placed into clean trays which are then transferred to a room tempered to approximately 100–115 F, where they are permitted to remain until they have grained.

Source: Crompton & Knowles

Grandma's Grained Molasses Mints

		% by Wt.
Granulated sugar	25 lb	86.21
Molasses (**Grandma's**)	4 lb	13.79
Water (to dissolve the sugar)		
Peppermint flavor		to suit
Red color		to suit
		Total 100.00

Procedure:

Place the sugar, molasses, and water into a kettle, stir occasionally and when the batch boils, wash down the sugar crystals that adhere to the kettle and boil the batch rapidly to 280–285 F (sea level).

Pour the batch on an oiled, cool slab, leaving in the kettle about $1^1/_2$ lb of the batch.

Replace the kettle with $1^1/_2$ lb of batch on the furnace, apply a little heat, then add about 4 oz of corn syrup and red color, mixing well. Cook the batch for about a minute, then pour this out to cool and use it for striping.

After the batch proper is moderately cool, fold it together and permit it to cool until fairly firm, then pull well on the hook, adding sufficient mint oil.

Form the batch into the shape of a loaf, then place the stripe in position and spin and cut the batch into pieces of the desired size and shape.

These candies will grain at normal temperatures within approximately an 8-h period. However, if it is desired to shorten the graining period, place the candies in a warm room at approximately 105–110 F for 3–4 h.

Source: Crompton & Knowles

MeloBan Creamettes

		% by Wt.
Sugar	9 lb	67.31
Nulomoline (invert sugar)	1 lb	7.48

Cream	1	qt	15.86
Water	1	qt	7.48
Nulco MeloBan (powdered ripe banana)	4	oz	1.87
			Total 100.00

Procedure:

Boil the sugar, **Nulomoline**, cream, and water to 275 F, pour on a slab to cool and when partially cold, color part of the batch brown for striping.

Add the **Nulco MeloBan**, mix well and pull thoroughly on the hook.

Place the stripes in position and spin and cut into pieces of the desired size. Dust the cut pieces well with powdered sugar and set aside until they grain through, much in the same manner as Creamed Mints.

Source: Crompton & Knowles

Nulofond Crystallized Creams

			% by Wt.
PART 1—FONDANT BASE:			
Nulofond (dry fondant)	50	lb	45.45
Water	5	lb	4.55
PART 2—BOB SYRUP:			
Granulated sugar	40	lb	36.36
Corn syrup	5	lb	4.55
Water	10	lb	9.09
			Total 100.00

Procedure:

PART 1—FONDANT BASE: Place ingredients into kettle and mix. When thoroughly mixed, set aside while preparing bob.

PART 2—BOB SYRUP: Place ingredients into kettle and cook to 242 F being careful to wash down all grains from the sides of the kettle.

Pour cooked bob syrup into the kettle containing fondant base, add flavor and color while mixing. Cast into starch or rubber molds. After creams have set, remove from molds. Immerse creams in crystallizing syrup cooled to room temperature; allow creams to remain in the crystallizing syrup until crystal coating has formed (5–10 h). Remove creams from crystallizing syrup, drain well and allow to dry before packing.

Source: Crompton & Knowles

Pecan Pralines

			% by Wt.
20% Butterfat cream	11	lb	12.10
Invert sugar	8.25	lb	9.08
Brownulated powdered brown sugar	8.25	lb	9.08
Salt	5	oz	0.34
Amerfond fondant sugar	50	lb	55.03
Chopped pecans	11	lb	12.10
Butter (melted)	1.5	lb	1.65
Invertase	3	oz	0.34
Imitation maple flavor	4	oz	0.28
			Total 100.00

Procedure:

Place cream, invert sugar, powdered brown sugar, and salt in remelt kettle or double boiler and heat, while stirring, to 175 F.

Add fondant sugar and stir until temperature reaches 160 F.

Add chopped pecans, melted butter, invertase, and maple flavor and mix well.

Spoon or deposit onto oiled paper or other release paper.

Source: Amstar

Pecan Roll

			% by Wt.
Brownulated powdered brown sugar	9.5	lb	8.97
Regular corn syrup	12.5	lb	11.81
Salt	6	oz	0.35
Water	6.25	lb	5.90
Amerfond fondant sugar	50	lb	47.23
Butter (melted)	6.25	lb	5.90

Vegetable fat (92 F m.p., melted)	6.25 lb	5.90
Chopped pecans	10 lb	9.45
Stock frappé	4.75 lb	4.49
		Total 100.00

Procedure:

Place powdered brown sugar, corn syrup, salt, and water in double boiler or remelt kettle and agitate, with heat, until sugar is dissolved (about 170 F).

Add fondant sugar and stir until smooth. Add melted butter and melted vegetable fat and mix well. Add pecans and frappé and mix well.

Pour at 160 F onto paper-lined table to desired height. Next day, cut into strips (preferably with wire cutting device). Dip in low cooked caramel and roll in pecans. Slice into $^1/_2$ in. pieces as desired.

Source: Amstar

Chapter 3

CHEWY CANDIES

As the name of this type of confection indicates, chewy candies are those that are chewable, i.e., malleable. They also possess the qualities of body and lubricity. The following are the most familiar varieties: caramels (nongrain), kisses (e.g., butterscotch kisses), jelly candies, taffy, and toffee.

Almondolas

			% by Wt.
Burnt sugar	1.00	lb	6.27
Powdered cream (well mixed in 8 lb water)	3.00	lb	18.82
Granulated sugar	5.00	lb	31.36
Regular corn syrup	5.00	lb	31.36
Salt	3.00	oz	1.18
Powdered mace	2.50	g	0.03
Butter	1.25	lb	7.84
Coffee-liquor chocolate*	8.00	oz	3.14
			Total 100.00

Procedure:

Mix burnt sugar, granulated sugar, mace, corn syrup, salt, and half of powdered cream mixture. Bring to a boil and add remaining powdered cream mixture slowly. Continue to cook and add coffee-liquor, chocolate, and butter. Cook to 252 F or medium ball. Deposit or funnel on a slab the size of half dollar. When cool, add two roasted Jordanola almonds to each patty. Roll, then wrap in cellophane.

Source: California Almond Growers Exchange

* This is liquor chocolate with pure coffee milled in and obtainable from several chocolate manufacturers.

Butter Mint Sticks

			% by Wt.
Granulated sugar	20	lb	74.07
Water	4	lb	14.81
Molasses	2	lb	7.41
Salted dairy butter	1	lb	3.71
Salt			to suit
Mint oil			to suit
			Total 100.00

Procedure:

Place all of the above ingredients into a kettle, stir occasionally until the batch boils, then cook to 275 F (at sea level).

Pour the batch out to cool, add the salt, and when the batch has cooled to approximately 125 F pull it for approximately 5 min on the power puller, adding the mint oil while pulling.

Spread the batch to form a flat piece, then place into the center of the batch approximately 8 lb of chewing caramel which has been slightly warmed to make it flexible. The purpose of the caramel is to form a core in the center of the candy.

Fold the batch around the caramel core, then spin into sticks which may later be cut into 4 in. or 6 in. lengths.

Permit the butter-mint sticks to grain before packing.

Source: Crompton & Knowles

Caramel

Formula No. 1

	% by Wt.
Part A	
Evaporated milk	37.20
Isomerose® 100 (high-fructose corn syrup)	30.00
Part B	
Vegetable fat (92–96 M.P.)	2.00
Clinton Special sweet corn syrup (52 D.E.)	30.00
Lecithin	0.05
Glycerol monostearate	0.15

Part C	
Salt	0.60
Vanilla	to suit
	Total 100.00

Procedure:

Blend Part A well. Add half of this mixture to Part B and bring to boil. Add remaining half of mixture and cook to 245 F. Add Part C as desired. Mix well.

NOTES:

This caramel is a soft tender product unusually well suited for bar goods and other coated items. Finished moisture will be in the range of 10.0–11.0%.

Source: ADM

No. 2

			% by Wt.
Corn syrup (regular)	2	lb	25.34
Granulated sugar	2	lb	25.34
Water (cold)	2	lb	25.34
Cream powder	1	lb	12.67
Salt	8	g	0.22
Fat	7	oz	5.54
Butter	7	oz	5.54
Butterscotch flavor			0.01
			Total 100.00

Procedure:

Whisk cream powder and water together for 3 min. Place other ingredients, except for flavor, in a boiling pan and add the cream suspension. Boil with continuous stirring to the firm ball stage. Add the flavor, mix thoroughly and pour on to silicone paper covered slab to a depth of $^3/_4$ in. When cold, cut into pieces 1 in. × $^3/_4$ in.

Coat top (only) with tempered milk chocolate, roll in **Bitsyn.** Then cover completely with milk chocolate.

Source: Pillsbury Co.

No. 3

Sugar	12.0	lb
Corn syrup	8.0	lb
Water	4.0	qt
Cream	8.5	qt
Hard butter	1.0	lb
Almonds	4.0	lb

Procedure:

Place sugar, glucose (corn syrup), and water in copper kettle and put on the fire, stirring until well dissolved. Stir and cook, adding the cream and hard butter gradually. Cook to a ball and set off, stir a little and then beat until the batch is quite creamy, then add almonds. Pour on greased slab between iron bars, allow to cool, and then cut into strips and again into oblongs.

Source: Savage Brothers Co.

Chocolate Caramels

			% by Wt.
Part A:			
Sugar	12	lb	58.50
Milk	4.5	qt	21.90
Butter	2	lb	9.80
Chocolate	2	lb	9.80
Lemon or vanilla			to suit
			Total 100.00
Part B:			
Sugar	10	lb	22.90
Corn syrup	20	lb	45.80
Cocoa butter	3	oz	0.40
Hard butter	13	oz	1.90
Cream	12.75	lb	29.00
			Total 100.00

Procedure:

Part A: Mix sugar, milk, and butter; set over a good fire and let the mixture come to a boil, then add chocolate, scraped fine. Let it boil until quite thick, stirring all the time. It will be better if the chocolate is mixed smooth with a pint or so of the boiling milk before adding the whole batch. When done, use lemon or vanilla to flavor, and pour out on slab between iron bars, about $^1/_2$ in. thick. When nearly cold, use the caramel cutter to finish.

Part B: Cook sugar, glucose (corn syrup), cocoa butter, hard butter, and $^1/_2$ of the cream to 275 F. Add other half of cream gradually and cook to first crack; set off the fire and add the cream batch which was first made, stir carefully until the cream is melted, and pour out on greased slab.

When cool enough to handle, knead like bread, place between iron bars, roll smooth and cut witht he caramel knife.

Source: Savage Brothers Co.

Economy Caramel

	% by Wt.
Nusweet "E" corn syrup (44° Baumé)	34.60
Sucrose	13.30
Nonfat milk solids	12.00
Water	18.60
Vegetable oil	12.00
Clinjel 260-B food starch (modified)	1.30
Water	8.00
Salt	0.20
Vanilla	to suit
	Total 100.00

Procedure:

Combine corn syrup and sucrose in steam-jacketed kettle. Combine nonfat milk solids and warm water under good agitation. Add to steam-jacketed kettle and apply heat. Add vegetable oil. Maintain agitation. Bring to boil. Suspend starch in cold water. Add slowly to boiling mixture, being sure not to interrupt boiling. Cook to desired temperature (approximately 250 F) or consistency. Add salt and flavor as desired.

Source: ADM

Caramels

(For Casting in Starch)

		% by Wt.
Stabilizer[1]		
Isomerose® 100 high fructose corn syrup	49.3 lb	35.55
Clinton regular corn syrup (43° Baumé)	50.0 lb	36.05
Granulated sugar	35 lb	25.23
Hardened vegetable fat	4 lb	2.88
Evaporated whole milk[2]		0.29
Salt	8 oz	
Vanilla flavor		to suit
		Total 100.00

Procedure:

Dissolve stabilizer in a small amount of water and stir this into the milk. Place **Isomerose® 100** brand high fructose corn syrup, regular corn syrup, sugar, vegetable fat, and approximately one-third of the milk in cooking kettle with an agitator. Mix and cook until the batch thickens. Add the remainder of milk so there is little or no interruption in boiling. Cook batch to desired consistency. In general, a caramel of the medium or medium-hard ball type is desired for starch casting. Add salt and flavor. Cast into starch. Remove from starch on following day and coat with chocolate.

NOTES:

1 1.25 oz ammonium carbonate or 1.25 oz sodium bicarbonate.

2 Sweetened condensed milk may be used in place of evaporated milk. If sweetened condensed milk is used, 21 lb of granulated sugar should be deducted from the formula, since this milk product contains an equivalent quantity of sugar. Approximately 20 lb water should also be added if sweetened condensed milk is used.

This formula cooked to a very soft ball stage (240–242 F) is also suitable for dipping caramel for coating marshmallow, nougat, etc.

Source: ADM

Caramel Almond Clings

			% by Wt.
Corn syrup	25	lb	20.45
Invert sugar	10	lb	8.18
Sweetened condensed whole milk	40	lb	32.72
Granulated sugar	15	lb	12.27
Dairy salted butter	5	lb	4.09
Hard coconut butter	2	lb	1.64
Salt	$^1/_4$	lb	0.20
Vanilla and maple flavors			to suit
Roasted almonds (select sheller run)	25	lb	20.45
			Total 100.00

Procedure:

Place all ingredients (excepting flavors) in caramel cooking kettle equipped with double-action, mechanical mixer-stirrers; mix well, apply heat, stir and cook to approximately 235–238 F (to obtain medium soft-ball consistency).

Add flavor, mix, and deposit caramel on metal or wooden trays covered to approximately $^1/_2$ in. deep with roasted almonds. When cool, cut into oblong shapes or squares. Enrobe or hand dip three-quarters of the way up, leaving the caramel almond surface exposed. Coat on the caramel surface or completely coat with chocolate.

(Condensed milk may be replaced with 40 lb of evaporated, unsweetened milk, adding 10 lb of granulated sugar. Start with 3 qt of unsweetened evaporated milk, adding balance gradually while boiling.)

Source: California Almond Growers Exchange

Caramel Whirls (Cut)

			% by Wt.
Condensed skimmed milk (40% sugar)	35	lb	24.07
Corn syrup	50	lb	34.39
Hard coconut butter (90 m.p.)	5	lb	3.44
Salt	6	oz	0.26
Nulomoline (invert sugar)	10	lb	6.88
Fine granulated sugar	20	lb	13.76
Soft winter wheat flour	10	lb	6.88

Fondant (80–20 cooked to 244 F)	15	lb	10.32
Flavor and color			to suit
			Total 100.00

Procedure:

Place the corn syrup, **Nulomoline**, milk and fat, also salt, into a kettle. Start the mixer in motion, turn the steam on partly; heat and mix until the fat is completely dissolved, then add sugar, turn steam on full and cook the batch to a hard ball. Turn the steam down so that just enough steam enters the kettle to keep the batch warm. Sift the flour into the batch and mix thoroughly. Spread batch out on a slab to cool.

Source: Crompton & Knowles

Cream Center for Caramel Wheels

			% by Wt.
Refined sugar	20	lb	27.78
Water	4	lb	5.56
Nulomoline (invert sugar)	5	lb	6.94
Corn syrup	10	lb	13.89
Fondant (80–20 cooked to 244 F)	25	lb	34.72
Standard Frappé (q.v.)	5	lb	6.94
Hard coconut butter	3	lb	4.17
			Total 100.00

Procedure:

Boil the sugar, water, corn syrup, and **Nulomoline** to 255 F. Turn off the heat, add the coconut butter and fondant, mix until fondant is dissolved, then add the frappé, flavor and color as desired, and mix thoroughly. Permit the batch to cool to approximately 150 F, then spread it on the sized caramel sheets.

STANDARD FRAPPÉ:

Nulomoline (invert sugar)	50	lb
Corn syrup	50	lb
Albumen (dissolved in water)	1	lb
Water	2	lb

Procedure:

First cook 50 lb of corn syrup to a temperature of 245 F. Shut off the steam and add 50 lb of cold **Nulomoline** and stir until the latter is melted. Place the batch into a marshmallow beater, start the machine, then add the albumen solution. Beat until light and use as needed.

More or less frappé may be used according to the weight of cream center desired. (For each 550 ft above sea level, reduce the boiling temperature by 1 F.)

Source: Crompton & Knowles

Cast Grained Caramels

(Medium priced bulk or package line)

		% by Wt.
Corn syrup	60 lb	37.39
Nulomoline (invert sugar)	10 lb	6.23
Coconut oil (76 F)	10 lb	6.23
Condensed milk (whole or skim)	40 lb	24.92
Basic fondant (q.v.)	40 lb	24.92
Salt	0.5 lb	0.31
Flavor		to suit
		Total 100.00

Procedure:

Place the corn syrup, Nulomoline, coconut oil, and condensed milk into a kettle and stir; cook to a medium firm ball, then add the salt, fondant, flavor and mix well. Cast the batch into starch.

These caramels may be chocolate coated or iced.

BASIC FONDANT:

Sugar	80 lb
Corn syrup	20 lb
Nulomoline (invert sugar)	10 lb
Water (to dissolve sugar)	

Procedure:

Cook to 244 F, cool and beat the fondant

Source: Crompton & Knowles

Special Semi-Short Chocolate Caramels (Cut)

			% by Wt.
Corn syrup	40	lb	25.74
Nulomoline (invert sugar)	10	lb	6.44
Skimmed condensed milk (40% sugar)	40	lb	25.74
Hard coconut butter (90 F m.p.)	5	lb	3.22
Cocoa powder (20–25% cocoa butter)	10	lb	6.44
Salt	6	oz	0.24
65–36 Fondant (cooked to 245 F)	50	lb	32.18
Flavor			to suit
			Total 100.00

Procedure:

Place the corn syrup, **Nulomoline**, and skimmed milk into a kettle, add fat and sift cocoa powder into the batch. Start the mixer in motion, turn steam on partly; after cocoa powder has been thoroughly mixed with other ingredients, turn steam on full and boil the batch to a hard crack. Shut off steam, add the fondant, salt, and flavor; mix well. Spread the batch out to cool.

Source: Crompton & Knowles

Cast Molasses Chewing Centers

			% by Wt.
Molasses	15	lb	21.39
Corn syrup	25	lb	35.65
Sugar	10	lb	14.26
Nulomoline (invert sugar)	5	lb	7.13
Dairy butter	3	lb	4.28
Coconut butter (76 F m.p.)	2	lb	2.85
Salt	2	oz	0.18
Special Frappé (q.v.)	10	lb	14.26
			Total 100.00

Procedure:

Place all above ingredients except the frappé into a kettle, stir and cook the batch to the first crack. Turn off heat, add frappé and mix thoroughly. Heat the hopper of the depositor to approximately 175 F, then cast the batch into starch and sprinkle the tops of the trays with starch. When the centers are cool they are ready to dip.

SPECIAL 50/50 FRAPPÉ:

Corn syrup	50 lb
Nulomoline (invert sugar)	50 lb
Albumen	2.5 lb
Water	5 lb

Procedure:

Boil corn syrup to 250 F, turn off the steam, add the **Nulomoline**, mix well, and place this into a marshmallow beater. Add the albumen which has been properly dissolved in the water and beat the batch until light. Set aside for future use.

Source: Crompton & Knowles

Chewy Nougat

		% by Wt.
Part A		
Dried egg albumen	2 oz	1.48
Water (cold)	7 oz	5.20
Powdered sugar	8 oz	5.94
Part B:		
Sugar (granulated)	2 lb	23.77
Water	12 oz	8.92
Corn syrup (regular)	3 lb	44.59
Part C:		
Fondant (4/1)	4 oz	2.97
Vanilla-vanillin (fourfold)		to suit
Fat	1.2 oz	0.95
Part D:		
Bitsyn® or nut pieces	8.5 oz	6.18
		Total 100.00

Procedure:

Place Part A water in bowl of Blakeslee mixer and run at lowest speed (wire beater). Add albumen slowly making sure all lumps are dispersed. Then add powdered sugar in the same manner. When albumen and sugar are fully dispersed, beat to a peak. Turn motor off, remover wire beater, and replace it with flat beater.

Boil Part B to 264 F, set beater to slow speed, and add syrup to Part A in a thin stream.

Add the ingredients of Part C in the order given. Do not overmix, especially after adding fat.

Weigh out lots of 3 lb and fold **Bitsyn®** into each.

Spread on cold table previously dusted with powdered sugar and level gently to $^1/_2$ in. bars. Dust with powdered sugar. When firm, cut into pieces $1^3/_8 \times {}^7/_8$ in. and wrap.

Source: Pillsbury Co.

Dipping Caramel

			% by Wt.
Granulated sugar	2.5	lb	20.61
Corn syrup (regular)	3	lb	27.49
Fat	1	lb	9.16
Sweetened condensed milk (whole)	4.25	lb	38.94
Salt	0.015	lb	0.14
Evaporated milk	0.40	lb	3.66
		Total	100.00

Procedure:

Cook to a full boil, then add evaporated milk in a fine stream, without stopping the boiling. Continue cooking to 242 F. Dip bars into caramel and roll in **Bitsyn®**. When set, cut and wrap.

Source: Pillsbury Co.

Full Cream Almond Lady Caraments

		% by Wt.
Corn syrup	15.5 lb	76.00
Sugar	13.5 lb	6.22
Cream (24%)	3 gal	11.74
Butter	2 lb	0.92
Roasted almonds (buttered-diced)	7.5 lb	3.45
Coconut oil (92°)	2 lb	0.92
Salt	2 oz	0.06
Promine "D"	1.5 lb	0.69
		Total 100.00

Procedure:

Cook sugar, corn syrup, cream, coconut oil, and butter to a soft ball stage—236 F. Add salt, almonds, and **Promine "D"**, mix well, pour on slab. This caramel holds its shape when cut, or may be wrapped in cellophane to prevent drying, or chocolate coated to add even greater shelf-life.

Source: California Almond Growers Exchange

Jelly Caramel for Chocolate Dipping

		% by Wt.
Water (cold)	60 lb	35.98
Cold boiling starch	6 lb	3.60
Dissolve in kettle and add:		
Yellow sugar	40 lb	23.99
Corn syrup	24 lb	14.39
Nulomoline	16 lb	9.60
Salt	6 oz	0.22
Cook to string same as gum and add:		
Condensed milk	15 lb	9.00
Continue boiling to good string and add:		
Vanilla	6 oz	0.22
Butter	5 lb	3.00
		Total 100.00

Procedure:

Cast in starch, place in drying room, leave overnight, let stand outside until cool and dip.

Source: Crompton & Knowles

Molasses Curly Cues

			% by Wt.
Thin boiling starch	3	lb	2.39
Water (cold)	3	lb	2.39
Corn syrup	65	lb	51.79
Molasses (**Grandma's**)	35	lb	27.89
Hardened coconut oil (96 F m.p.)	5	lb	3.98
Salt	8	oz	0.40
Peppermint, butter, cinnamon, or other flavors			to suit
Frappé No. 1 (q.v.)	10	lb	7.97
Dutched cocoa powder	4	lb	3.19
			Total 100.00

Procedure:

Mix the starch with 3 lb of cold water. Add corn syrup, molasses, and hardened coconut oil and bring batch to boiling point. Continue to stir and cook to 260 F (sea level), then add the salt, peppermint or other flavors and mix well. Now add frappé and mix thoroughly. Spread half of the batch as thinly as possible on an oiled slab.

Sift cocoa powder into remainder of the batch, mixing well, then spread this portion of the batch on top of the thinly spread batch on slab.

When the batch is almost cool, roll it in much the same manner as for jelly rolls, then spin, size, and cut into small pieces which may be wrapped in waxed paper, cellophane or coated with chocolate.

FRAPPÉ NO. 1:

Corn syrup	50	lb
Invert sugar	50	lb
Albumen	1	lb
Water	2	lb

Procedure:

Cook corn syrup to 245 F (sea level), shut off the heat and add invert sugar and stir until it is melted. Place the batch into a marshmallow beater, start the beater, then gradually add albumen which has been previously dissolved in water and beat the batch until light.

Source: Crompton & Knowles

Nongraining Dipping Caramel

			% by Wt.
Corn syrup	40	lb	36.28
Nulomoline (invert sugar)	20	lb	18.14
Fine granulated sugar	10	lb	9.07
Condensed milk (whole or skim)	30	lb	27.21
Coconut butter	10	lb	9.07
Fine table salt	4	oz	0.23
			Total 100.00

Procedure:

Place the coconut butter, corn syrup, **Nulomoline**, and condensed milk into a kettle. Heat all together until well mixed; then add granulated sugar; boil batch to the desired consistency. Add salt and flavor, mix well, and place into reservoir of the coating machine.

Source: Crompton & Knowles

Nut Caramel Tops

(Caramel Deposited onto Nut Meats)

			% by Wt.
Granulated sugar	3	lb	15.28
Corn syrup	5	lb	25.49
Dairy salt butter	1	lb	4.08
Hardened coconut oil (92–100 F m.p.)	1	lb	4.08
Nulomoline (invert sugar)	2	lb	10.19

Sweetened condensed whole milk	8	lb	40.76
Salt	1	oz	0.12
			Total 100.00

Procedure:

Place all the ingredients in a caramel cooking kettle, apply heat, mix well, and cook the batch to 235–237 F. Add either vanilla, orange, almond, or maple flavor. Mix thoroughly and deposit the caramel onto toasted, slivered almonds, pecans, or other nut meats which are placed in trays or pans.

Source: Crompton & Knowles

Nutty Chews

			% by Wt.
Corn syrup	50	lb	32.19
Molasses (**Grandma's**)	20	lb	12.88
Nulomoline	5	lb	3.22
Coconut oil	5	lb	3.22
Basic fondant (q.v.)	25	lb	16.10
Salt	5	oz	0.20
Roughly ground or roasted broken peanuts	50	lb	32.19
			Total 100.00

Procedure:

Boil the corn syrup, molasses, **Nulomoline**, and coconut oil to 242–248 F. Add the fondant and stir until melted. Work in the salt and peanuts, spread the batch out on an oiled and dusted slab and when cold, size and cut into pieces, then coat in icing or chocolate, or wrap in waxed paper.

BASIC FONDANT:

Granulated sugar	80	lb
Corn syrup	20	lb
Invert sugar	10	lb
Water		

Procedure:

Boil granulated sugar, corn syrup, and invert sugar, plus sufficient water to dissolve the sugar, to 242 F (sea level), then cool and cream into fondant.

Source: Crompton & Knowles

Paste Caramels

			% by Wt.
Caramel paste	20	lb	95.00
Cocoa butter	1	lb	5.00
Vanilla or chocolate			to suit
			Total 100.00

Procedure:

Cook caramel paste and cocoa butter to a very stiff ball, in a copper kettle. Add nuts of any sort, well broken, rather than chopped, and flavor with vanilla or chocolate, and it is ready for the slab.

Source: Savage Brothers Co.

Pecan Roll

			% by Wt.
Brown sugar	9.5	oz	8.64
Corn syrup (regular)	12.5	oz	11.36
Salt	0.5	oz	0.45
Water	6.25	oz	5.68
Dry fondant (sifted)	50	oz	45.46
Salted dairy butter (melted)	6	oz	5.45
Melted fat	6.5	oz	5.91
Bitsyn® or pecan pieces	14	oz	12.73
Frappé	4.75	oz	4.32
			Total 100.00

Procedure:

Place brown sugar, corn syrup, salt, and water in a double boiler and agitate until sugar is dissolved (ca. 170 F). Add dry fondant and stir until smooth. Add melted butter and fat and mix. Add **Bitsyn**® and frappé and mix. Pour at 160 F on to paper and level to $^3/_4$-in. depth. When cool, cut into bars 4 × $^7/_8$ in. on wire cutter.

Source: Pillsbury Co.

Philadelphia Caramel Clinkers

			% by Wt.
Hardened edible oil (86–96 m.p.)	7	lb	7.18
Sweetened condensed nonfat milk	40	lb	41.03
Corn syrup	25	lb	25.64
Granulated sugar	15	lb	15.38
Nulomoline (invert sugar)	10	lb	10.26
Salt	8	oz	0.51
Vanilla or maple flavor			to suit
			Total 100.00

Procedure:

Place all of the above ingredients, except the flavor, into a caramel cooking kettle; apply heat, mix, and cook to approximately 232 F. Add the color and flavor, then transfer the batch to the hopper of a depositor and deposit the caramel onto pecans, almonds, or other nut meats.

The above produces relatively smooth caramel but if additional body is required, it is suggested that approximately 4 lb of soft winter wheat be first mixed with the hardened edible oil and then cooked with the batch. The flour may cause slight roughness of texture; if this is not desirable, increase the amount of condensed milk by 20% of its weight.

Source: Crompton & Knowles

Plastic Cast Caramels

			% by Wt.
Sugar	45	lb	21.34
Corn syrup	35	lb	16.58
Nulomoline (invert sugar)	10	lb	4.74
Starch (#50–60)	8	lb	3.79
Water	70	lb	33.13
Cream of tartar	2	oz	0.06

Boil all together to a good jelly (220–222 F), then add:

Condensed milk (40% sugar)	40	lb	18.94
Coconut oil (92 m.p.)	3	lb	1.42
Salt and flavor			to suit
			Total 100.00

Procedure:

Cook to a good jelly (84 Brix); cast batch into dry starch. Place the trays into dry room (temp. 130 F) for 15 h. When caramels have set, remove from starch; coat in icing or chocolate. For a good bar combination, fill the molds one-third with caramel and the balance with marshmallow.

Source: Crompton & Knowles

Quality Milk Caramels

(Pack in Parchment Cups, Wrap, or Coat)

			% by Wt.
Coconut oil (86–96 m.p.)	2	lb	4.53
Lecithin	1/2	oz	0.11
Granulated sugar	4	lb	9.06
Corn syrup	12	lb	27.18
Nulomoline (invert sugar)	4	lb	13.59
Sweetened condensed whole or skimmed milk	20	lb	45.30
Salt	1	oz	0.23
Maple, vanilla, coffee, or other flavors			to suit
			Total 100.00

Procedure:

Place the coconut oil in a caramel cooking kettle equipped with double-action stirrers and heat until melted. Add the lecithin and mix thoroughly.

Now add the remainder of the ingredients, except the flavor, and cook to a medium-firm ball stirring constantly. Add the flavor and nut meats, if desired, and mix thoroughly.

Spread the batch out to cool, leveling off to the required height. Cut into squares or oblongs and pack into parchment paper cups, wrap in waxed paper or cellophane, or coat with chocolate.

Source: Crompton & Knowles

Semichewing (Grained) Caramels

			% by Wt.
Corn syrup	10	lb	21.39
Nulomoline (invert sugar)	1.5	lb	3.21

Sweetened condensed whole milk	20	lb	42.78
Salt	4	oz	0.53
Dairy butter or hardened vegetable oil	3	lb	6.42
Short Fondant	7	lb	14.97
Frappé No. 2 (q.v.)	5	lb	10.70
Vanilla, maple, rum or other flavor			to suit
			Total 100.00

Procedure:

Place all the above ingredients, except the fondant, frappé, and flavor into a standard caramel cooking kettle equipped with agitators. Stir and cook the batch to approximately 250 F. Permit the batch to cool to 180–190 F, then add the fondant, mixing until the fondant is melted. Add the frappé and flavor, spreading batch out in much the same manner as standard chewing caramels. When the batch is still slightly warm (about 105 F) it may be spun into strips or fed into a kiss-wrapping machine, sized and cut, then wrapped in cellophane to form kisses.

The sheets of grained caramels may be cut into squares or oblongs and, of course, the variety can be further extended by adding whole or broken nut meats or by changing the color and flavor.

A grained caramel of this type may be rolled down to form a sheet approximately $^3/_8$ in. thick, then used as a layer for molasses coconut chewing centers, cutting the batch into pieces of the desired size and shape.

FRAPPÉ NO. 2:

Corn syrup	25	lb
Nulomoline (invert sugar)	25	lb
Albumen (dissolved in the water)	1	lb
Water (cold)	2	lb

Procedure:

Boil the corn syrup to 245 F, add the **Nulomoline** and mix until it is completely dissolved. Turn off the heat, place the batch into a beating machine, start the beater, then gradually add the albumen solution, and beat the batch until it is quite stiff. Remove the frappé from the beater and place into a suitable container for future use.

Source: Crompton & Knowles

Special Semishort Chocolate Caramels

(Cut)

			% by Wt.
Corn syrup	40	lb	25.74
Nulomoline (invert sugar)	10	lb	6.44
Skimmed condensed milk (40% sugar)	40	lb	25.74
Hard coconut butter (90 F m.p.)	5	lb	3.22
Cocoa powder (20–25% cocoa butter)	10	lb	6.44
Salt	6	oz	0.24
Fondant 65–35 (cooked to 245 F)	50	lb	32.18
Flavor			to suit
			Total 100.00

Procedure:

Place the corn syrup, **Nulomoline**, and skimmed milk into a kettle, add the fat and sift the cocoa powder into the batch. Start the mixer in motion, turn the steam on partly and after the cocoa powder has been thoroughly mixed with the other ingredients, turn the steam on full and boil the batch to a hard crack. Shut off the steam, add the fondant, salt, and flavor and mix well. Spread the batch out to cool and handle in the usual manner.

Source: Crompton & Knowles

Summer Chocolate Caramel

			% by Wt.
Granulated sugar	8	lb	20.51
Nulomoline (invert sugar)	2	lb	5.13
Corn syrup	10	lb	25.64
Evaporated unsweetened milk	15	lb	38.47
Dairy salt butter	1.5	lb	3.84
Chocolate liquor	2.5	lb	6.41
Salt and vanilla or other flavors			to suit
			Total 100.00

Procedure:

Place the sugar, **Nulomoline**, corn syrup, butter, chocolate liquor, and approximately one-half of the evaporated milk into a kettle. Stir occasionally until it boils, then gradually add the remainder of the evaporated milk, a little at a time, keeping the batch constantly boiling. As the batch becomes thick, stir rapidly to prevent scorching. Cook to a medium-firm ball, turn off the heat, and add the salt, vanilla, or other flavors. Mix well.

Spread the batch out to set, leveling it off to the required height. When cold, cut it into oblongs or squares and place into parchment paper cups.

Source: Crompton & Knowles

Summer Vanilla Caramels

(Unwrapped)

			% by Wt.
Granulated sugar	8	lb	19.05
Nulomoline (invert sugar)	2	lb	4.76
Corn syrup	10	lb	23.81
Dairy salt butter	2	lb	4.76
Evaporated unsweetened milk	20	lb	47.62
Salt, vanilla, coffee, maple, rum, raspberry, or pistachio flavors, with colors to match			to suit
			Total 100.00

Procedure:

Place the sugar, **Nulomoline**, corn syrup, butter, and one-half of the evaporated milk into a kettle, and mix well. Bring the batch to a boil, stirring occasionally. While boiling, gradually add the remaining evaporated milk a little at a time, stirring constantly and rapidly. Cook to a medium-firm ball; then add the salt, vanilla, or other flavors.

Spread the batch out to the required height (thus avoiding sizing). When cold, cut into squares, oblongs, etc. If desired, raw or roasted almonds or Brazil nuts may be added.

NOTE:

A richer caramel may be produced by replacing as much as half of the evaporated milk with cream of 22–36% butterfat.

Source: Crompton & Knowles

"Superfine" Cream Caramels

			% by Wt.
Corn syrup	7	lb	22.95
Sugar	9	lb	29.51
Cream (20% butterfat)	8.5	lb	27.87
Cream, condensed	6	lb	19.67
			Total 100.00

Procedure:

Mix glucose (corn syrup), sugar, 20% cream, heavy condensed cream. Cook about 242 F in winter and 244 F in summer, or to a medium ball when tested in cold water.

Start with a slow fire, boil slowly and stir constantly. Add vanilla flavor just before pouring on the slab. When nearly cold, mark caramel size, and cut; they will not get hard, but will be pliable without being soft.

Source: Savage Brothers Co.

Wrapped Caramels

(Cut or Cast)

			% by Wt.
Corn syrup	25	lb	24.98
Nulomoline (invert sugar)	5	lb	5.00
Cream (18% butterfat)	20	lb	19.99
Granulated sugar	22	lb	21.99
Hardened edible oil	3	lb	3.00
Evaporated milk	25	lb	24.98
Salt	1	oz	0.06
Flavor			to suit
			Total 100.00

Procedure:

Place the corn syrup, **Nulomoline**, and 30% of the cream into a caramel cooking kettle, add the sugar and edible oil and bring the batch to the boiling point.

Meanwhile, mix balance of cream with evaporated milk and gradually add this to batch while boiling; cook to firm ball. Add the salt and flavor, as desired, mix well, then spread the batch out to cool, size, cut, and wrap.

This formula may be used for caramels that are to be coated with chocolate, or batch may be rolled into thin sheets $^1/_4$ in. thick and used as a layer on short or chewy nougat, or on chewy coconut work

Source: Crompton & Knowles

Almond Divinity Kisses

			% by Wt.
Egg albumen	4	oz	0.98
Water	11	oz	2.69
Sugar	13	lb	50.90
Corn syrup	5	lb	19.57
Fondant 80/20	11	oz	2.69
Invert sugar	1	lb	5.19
Salt	1.5	oz	0.36
Select sheller run almonds, roasted and chopped	4.5	lb	17.62
Vanilla flavor			to suit
			Total 100.00

Procedure:

Dissolve egg albumen with the water. Place the invert sugar and egg albumen in Hobart beater. Beat until stiff. Cook the sugar and corn syrup to 240 F. Add the cooked syrup in a very fine stream into the Hobart beater. Continue beating, add the fondant and salt. Continue beating until batch shows a slight grain. Then fold in the almonds and flavor. Remove from beater and spoon out into desired shape and size.

Source: California Almond Growers Exchange

Butterscotch Kisses

			% by Wt.
Corn syrup	30	lb	34.68
Nulomoline (invert sugar)	10	lb	11.56
Granulated sugar	20	lb	23.12
Condensed whole milk	10	lb	11.56

Corn starch	1 lb	1.16
Water	5 lb	5.78
Salted dairy butter	2 lb	2.31
Hardened edible oil	3 lb	3.47
Frappé No. 1 (q.v.)	5 lb	5.78
Salt	0.5 lb	0.58
Vanilla or butterscotch flavor		to suit
		Total 100.00

Procedure:

Place the corn syrup, **Nulomoline**, condensed milk, sugar, corn starch, and water, dairy butter, and coconut oil into a kettle, mix well, apply heat, and cook the batch to a firm ball. Then add the salt, frappé, and flavor. Mix well; pour the batch out to cool and when the batch is moderately plastic, feed it to the wrapping machine.

FRAPPÉ NO. 1:

Corn syrup	50 lb
Invert sugar	50 lb
Albumen	1 lb
Water	2 lb

Procedure:

Cook corn syrup to 245 F (sea level), shut off the heat and add invert sugar and stir until it is melted. Place the batch into a marshmallow beater, start the beater, then gradually add albumen which has been previously dissolved in water and beat the batch until light.

Source: Crompton & Knowles

Seashore Wrapping Taffy Kisses

		% by Wt.
Corn syrup	40 lb	44.35
Granulated sugar	30 lb	33.26
Nulomoline (invert sugar)	10 lb	11.09
Water	6 lb	6.65

Hardened edible oil (86–96 F m.p.)	4	lb	4.44
Salt	3	oz	0.21
			Total 100.00

Procedure:

Place all of the above ingredients into a kettle, bring the batch to the boiling point and cook the batch to approximately 255–265 F (according to weather conditions), then pour the batch out to cool, and when partly cool, add color and flavor, pull well on the hook, then spin, cut and wrap.

If a richer taffy is desired, then cook the batch to 270–272 F, then add 8 lb sweetened condensed whole milk, mix well and cook the batch to the desired consistency.

Source: Crompton & Knowles

Superior Molasses Chewing Kisses

			% by Wt.
Corn syrup	25	lb	40.98
Molasses	15	lb	24.59
Nulomoline (invert sugar)	5	lb	8.20
Granulated sugar	10	lb	16.39
Salted dairy butter	2	lb	3.28
Hardened edible oil	2	lb	3.28
Water	2	lb	3.28
Vanilla, lemon, or orange flavor			to suit
			Total 100.00

Procedure:

Place all of the above ingredients into a kettle, mix well, stir occasionally until the batch boils and cook the batch to 256–260 F, then turn the batch out to cool.

When the batch is partially cool, pull well on the hook adding flavor, then spin, cut and wrap.

Source: Crompton & Knowles

Tomato Starch Jellies

			% by Wt.
Sugar	50	lb	24.03
Nulomoline (invert sugar)	25	lb	12.02
Corn syrup	25	lb	12.02
Starch	10	lb	4.81
Water	40	lb	19.21
Cream of tartar	1	oz	0.03
Tomato pulp and juice	7	gal	27.88
			Total 100.00

Procedure:

Rub the tomato pulp through a #18 sieve so as to remove the seeds. Place the tomato juice and strained pulp into a kettle, add the sugar, **Nulomoline**, corn syrup, water, and starch and heat to the boiling point. Add the cream of tartar and continue to boil and stir the batch to a good jelly.

Cast the batch into warm dry starch, sprinkle the surface of the jelly with starch and place in a drying room for 24 h at a temperature of approximately 125–130 F, then remove the jellies from the drying room and when cold, sand them in the usual way

Source: Crompton & Knowles

Agar Jellies

(Cut or Cast)

			% by Wt.
Agar no. 1	5	oz	0.93
Water	12	lb	37.15
Granulated sugar	10	lb	30.96
Corn syrup	5	lb	15.48
Nulomoline (invert sugar)	5	lb	15.48
Color and flavor			to suit
			Total 100.00

Procedure:

Soak the agar in the water for 1 h or more. Heat and boil the batch for 1–2 min to completely dissolve the agar. Add the granulated sugar and

Nulomoline, mix, and cook the batch to 222 F (at sea level); add the corn syrup, mixing well.

When adding citric acid to obtain a tart flavor, permit the batch to cool to approximately 125–120 F, then add the flavor, color, and citric acid solution, mixing well. Grated lemon or orange peel may also be added.

Cast the batch into dry starch impressions, sprinkling the surface of the cast jellies with dry starch; or, pour the batch into a tray lined with brown wrapping paper, sifting granulated or powdered sugar over the surface. When the jelly has set, remove the wrapping paper by moistening it with water, dust well with granulated or powdered sugar, then cut the jelly into squares, oblongs, etc.; then rolling in sugar. Permit the jellies to dry on the surface before packing.

Source: Crompton & Knowles

Agar (Japanese) Jellies

(Cut or Cast)

		% by Wt.
Agar agar (Japanese gelatin)	1.75 lb	0.99
Water	75 lb	42.44
Granulated sugar	50 lb	28.29
Corn syrup	25 lb	14.14
Nulomoline (invert sugar)	25 lb	14.14
Color and flavor		to suit
		Total 100.00

Procedure:

Soak the agar in the water for 1 h or more. Heat and boil the batch for 1–2 min to completely dissolve the agar, then add the granulated sugar, mix, and cool the batch to 222 F (at sea level); then add the **Nulomoline** and corn syrup, mixing well.

When adding citric acid to obtain a tart flavor, permit the batch to cool to approximately 125 F, then add the flavor, color, and citric acid solution, mixing well.

Cast the batch into dry starch impressions sprinkling the surface of the batch into a tray, which has been first lined with brown wrapping paper, sifting granulated sugar or powdered sugar over the surface. When the jelly has set, remove the wrapping paper by moistening it with water; dust well with granulated or powdered sugar, then cut the jelly into squares, oblongs, etc., rolling in sugar. Permit the jellies to dry on the surface before packing.

Source: Crompton & Knowles

All-Sugar Sanded Jellies

			% by Wt.
Granulated sugar	55	lb	25.91
Nulomoline (invert sugar)	45	lb	21.21
Thin-boiling starch #65	12	lb	5.66
Water	100	lb	47.13
Cream of tartar	3	oz	0.09
Color and flavor			to suit
			Total 100.00

Procedure:

Place the sugar, **Nulomoline**, water, starch, and cream of tartar into a kettle and cook. Stir the batch occasionally until it reaches the boiling point. Continue to stir and boil the batch to a good string. Turn off the heat, add color and flavor and cast the batch into dry starch.

Place the jellies in a drying room overnight, then remove the jellies from the drying room and clean well. Run them through a sanding machine and when the jellies are quite dry they are ready to be packed.

Source: Crompton & Knowles

Cast or Cut Pectin Jelly

			% by Wt.
Water	20	lb	32.32
Granulated sugar	30	lb	48.49
Citrus pectin-#100 grade	12	oz	1.21
Sodium acetate (U.S.P. or F.C.C.*)	1	oz	1.62
Nulomoline (invert sugar)	10	lb	16.16
Powdered citric acid	2	oz	0.20
Brazil nuts (whole, broken, sliced or ground)			to suit
Flavor and color			to suit
			Total 100.00

* F.C.C. = Food Chemicals Codex purity

Procedure:

Place the water into a kettle. Meanwhile mix the pectin with $^1/_3$ of the sugar. While heating the water, sift in the sugar pectin mixture, mixing well. Bring the batch to the boiling point and permit it to boil for a minute.

Dissolve the sodium acetate in a little water and add this to the batch, along with the **Nulomoline.**

Dissolve the citric acid in a little water, add half of the resulting citric acid solution and bring the batch to the boiling point.

Gradually add the balance of the sugar and cook the batch to 225–228 F (sea level), according to the consistency desired. Add flavor and color and the remainder of the citric acid solution and mix well.

Cast a portion of the jelly into moderately large starch impressions, then immediately drop medium size or large size whole Brazil nuts on top of the jelly, and shortly thereafter cast additional jelly into the starch impressions to cover the Brazil nuts.

Source: Crompton & Knowles

Convention Starch Jellies

			% by Wt.
Granulated sugar	55	lb	24.26
Nulomoline (invert sugar)	10	lb	4.41
Corn syrup	40	lb	17.64
Thin boiling starch #50–60	11.5	lb	5.07
Water	110	lb	48.54
Cream of tartar	3	oz	0.08
Orange flavor and orange color			to suit
			Total 100.00

Procedure:

Place the sugar, **Nulomoline,** $^1/_4$ of the corn syrup, starch, water, and cream of tartar into a kettle and gradually bring to the boiling point. Turn the steam on full and cook the batch as rapidly as possible to a sheet, then add the balance of the corn syrup, color, and flavor and mix well. Cast the batch into warm starch (110–130 F) and allow the jellies to remain in the starch for approximately 2 days, depending upon the size of the piece.

Source: Crompton & Knowles

Iced Flowing Jellies

(Cast)

			% by Wt.
Granulated sugar	65	lb	34.30
Nulomoline (invert sugar)	5	lb	2.64
Corn syrup	30	lb	15.83
Thin boiling starch #65	9.5	lb	5.01
Water	80	lb	42.22
Color and flavor			to suit
			Total 100.00

Procedure:

Place the sugar, **Nulomoline**, corn syrup, starch, and water into a kettle and bring the batch to the boiling point, then boil to a string. Cast the batch into warm dry starch and place into a drying room at a temperature of approximately 125 F for 12 h. Remove the jellies from the drying room and allow to set until cool, then clean them well and coat with icing.

Soucce: Crompton & Knowles

Fruit Jelly Bars

			% by Wt.
Granulated sugar	10	lb	31.34
Corn syrup	5	lb	15.67
Nulomoline (invert sugar)	5	lb	14.44
Agar No. 1	5	oz	0.94
Water	12	lb	37.61
Lemon, orange, etc. flavors			to suit
Preserved fruits, diced, whole, pineapple wedges, cherries, etc.			to suit
			Total 100.00

Procedure:

Cover a table top or slab with brown wrapping paper, place metal bars in position, spread sufficient preserved fruits over the paper. Cook the jelly batch as follows: mix the agar and water, apply heat and slowly bring to a boil with stirring, add the sugar and **Nulomoline** and cook to 222–223 F. Remove from heat, add the corn syrup, mix well, and then add the flavor. After mixing thoroughly, pour over the preserved fruit. Sprinkle desiccated coconut over the surface. When set, cut into bars, squares, etc.

Source: Crompton & Knowles

Hard Gum Drops with Gum Arabic

		% by Wt.
Granular gum arabic	10 lb	26.67
Warm water	10 lb	26.67

Prepare the above solution by gently stirring the gum arabic into the warm water. Allow to set overnight. Skim and filter the following day.
Combine the following in the steam-jacketed laboratory kettle:

Filtered gum solution	10 lb	26.67
Granulated sugar	3.5 lb	9.33
Water	1.5 lb	4.00
Corn syrup (43% Baumé)	2.5 lb	6.66
		Total 100.00

Procedure:

Turn on steam—cook cautiously to a temperature of 225 F. If batch foams excessively, add a few drops of vegetable oil. Check solids at end point of cook with refractometer. This should be approximately 75%. Cut off steam; add $^1/_2$ oz flavor and color; mix and cast into molding starch impressions. Allow to stand overnight, preferably in a hot room. Examine the following morning.

Source: Penwest Ltd.

Iced Flowing Jellies

(Cast)

		% by Wt.
Granulated sugar	65 lb	34.30
Nulomoline (invert sugar)	5 lb	2.64

Corn syrup	30 lb	15.83
Thin-boiling starch #65	9.5 lb	5.01
Water	80 lb	42.22
Color and flavor		to suit
		Total 100.00

Procedure:

Place the sugar, **Nulomoline**, corn syrup, starch, and water into a kettle and bring the batch to the boiling point, then boil to a string. Cast the batch into warm dry starch and place into a drying room at a temperature of approximately 125 F for 12 h. Remove the jellies from the drying room and allow to set until cool, then clean them well and coating with icing.

Source: Crompton & Knowles

Japanese Jellies for Cutting or Casting

		% by Wt.
Sugar	50 lb	28.30
Corn syrup	25 lb	14.14
Nulomoline (invert sugar)	25 lb	14.14
Japanese Gelatine #1	1.75 lb	0.99
Water	75 lb	42.43
		Total 100.00

Procedure:

Soak the **Japanese gelatine** in the water for at least 2 h. Place the soaked gelatin into a kettle, heat, stirring occasionally, and permit to boil for several minutes, then strain through a # 30 sieve.

Add the sugar and **Nulomoline** to the strained gelatin and boil the batch to 224 F. Turn off the heat, add the corn syrup, mix well, and when the batch has cooled to approximately 125 F, add flavor and color. The batch may now be poured into trays that have been lined with wrapping paper or cast into warm dry starch.

If the jelly is poured into trays, allow it to remain undisturbed until it has set, then turn the jelly sheet upside down. Moisten the paper with water, peel off the paper, and cut and roll the jellies in XXX sugar, then roll again in XXX sugar and permit the pieces to dry before they are packed into containers.

Source: Crompton & Knowles

Jelly Drops, Orange Jelly Slices, Etc.

			% by Wt.
Granulated sugar	40	lb	18.99
Nulomoline (invert sugar)	10	lb	4.75
Corn syrup	50	lb	23.74
Corn starch (65 fluidity)	10.5	lb	4.98
Cold water	100	lb	47.48
Cream of tartar	2	oz	0.06
			Total 100.00

Procedure:

Place the water, sugar, starch, and $^3/_5$ of the corn syrup into a kettle equipped with a good stirring device. Turn the steam on partly and allow the batch to gradually reach the boiling point, then add the cream of tartar, turn on steam on full and boil the batch as rapidly as possible using 80–100 lb steam pressure. Cook the batch to a string, turn off the heat, add the balance of the corn syrup and the **Nulomoline**, then color and flavor, mix thoroughly and cast the batch into starch.

The temperature of the starch into which the jellies are cast may range from 90–175 F.

If a continuous starch drier is available heat the starch to 160–175 F, then cast the batch into the starch and leave the jellies in a room at normal temperature for at least 10 h. They may then be removed from the starch, moistened with steam vapor and sanded with fine granulated sugar.

In the absence of a continuous starch drier the jellies may be placed in a drying room heated to 140 F and allowed to dry from 24–48 h.

NOTE:

For each 550 ft above sea level, reduce the boiling temperature by 1 F.

Source: Crompton & Knowles

Marmajell

(Cast Starch Jellies)

			% by Wt.
Granulated sugar	60	lb	27.15
Molasses	15	lb	6.79

Nulomoline (invert sugar)	25	lb	11.31
Starch (#40 fluidity)	11	lb	4.98
Water	90	lb	40.73
Frappé No. 1 (q.v.)	10	lb	4.52
Bleached raisins	10	lb	4.52
			Total 100.00

Procedure:

Place the sugar, molasses, and **Nulomoline** into a kettle. Mix the starch and water add to the sugar–molasses–**Nulomoline** mixture. Turn on the steam and cook to a heavy jelly body. Shut off the steam, add the raisins (ground), mix these through the batch, then add the frappé and mix again. Add red color and rum flavor.

Cast the batch into dry, warm starch and leave the goods in a dry room having a temperature of 130 F for 24 h, then remove and leave them in a room of ordinary temperature for approximately 24 h.

After the goods are taken from the starch sand them with sugar that has been colored to match the jellies.

FRAPPÉ NO. 1:

Corn syrup	50	lb
Invert sugar	50	lb
Albumen	1	lb
Water	2	lb

Procedure:

Cook corn syrup to 245 F (sea level), shut off the heat and add invert sugar and stir until it is melted. Place the batch into a marshmallow beater, start the beater, then gradually add albumen which has been previously dissolved in water and beat the batch until light.

Source: Crompton & Knowles

All-Sugar Sanded Jellies

			% by Wt.
Granulated sugar	55	lb	25.92
Nulomoline (invert sugar)	45	lb	21.21

Thin boiling starch #65	12	lb	5.66
Water	100	lb	47.12
Cream of tartar	3	oz	0.09
Color and flavor			to suit
			Total 100.00

Procedure:

Place the sugar, **Nulomoline**, water, starch, and cream of tartar into a kettle and cook and stir the batch occasionally until it reaches the boiling point. Continue to stir and boil the batch to a good string. Turn off the heat, add color and flavor and cast the batch into dry starch.

Place the jellies in a drying room overnight, then remove the jellies from the drying room and clean well. Run them through a sanding machine and when the jellies are quite dry they are ready to be packed.

Source: Crompton & Knowles

Nulo Super-Sweet Cast Starch Jelly Gums

			% by Wt.
Corn syrup	65	lb	34.20
Nulomoline (invert sugar)	35	lb	18.41
Thin boiling starch (#50–60)	10	lb	5.26
Water	80	lb	42.10
Citric or tartaric acid	1	oz	0.03
Color and flavor			to suit
			Total 100.00

Procedure:

Place the corn syrup and **Nulomoline** into a kettle. Mix the starch with part of the water, adding this along with the remainder of the water and acid, to the corn syrup–**Nulomoline** batch. Bring the batch to the boiling point, cooking to a good jelly string. Add color and flavor and cast the batch into dry, warm starch.

Allow the jellies to remain in the starch for approximately 48 h—or place into a drying room tempered to 130–135 F for a minimum of 24 h. Remove the jellies from the starch and sand, or coat with hardened vegetable oil of 98 F m.p., then rolling in powdered skim milk.

Source: Crompton & Knowles

Molasses Peanut Chews

(Cut)

			% by Wt.
Corn syrup	50	lb	33.27
Molasses (**Grandma's**)	15	lb	9.98
Nulomoline (invert sugar)	25	lb	16.63
Peanut butter	20	lb	13.30
Salt	5	oz	0.21
Roughly ground roasted peanuts	40	lb	26.61
		Total	100.00

Procedure:

Boil corn syrup, molasses, and **Nulomoline** to 240–244 F. Add the peanut butter and salt, mix thoroughly. Then work in ground roasted peanuts and spread the batch on an oiled and dusted slab, and when cold, size and cut into bars or smaller pieces. Then, coat in chocolate or icing.

Source: Crompton & Knowles

Wrapped Molasses Chews

(Suitable for Coating in Chocolate)

			% by Wt.
Molasses (**Grandma's**)	40	lb	24.92
Fine granulated sugar	30	lb	18.69
Corn syrup	50	lb	31.16
Hard coconut butter (88 F m.p.)	10	lb	6.23
Special Frappé (q.v.)	30	lb	18.69
Salt	8	oz	0.31
		Total	100.00

Procedure:

Place the corn syrup, molasses, and coconut butter into a kettle. Heat all together until coconut butter has been thoroughly mixed with other ingredients. Add the fine granulated sugar, mix well, then turn steam on to limit and boil batch as rapidly as possible to a good crack. Now turn off heat and add the frappé and salt.

After making one or two batches it will be possible to determine the proper temperature of boiled portion of batch so that when the frappé is mixed with the batch, it will not be necessary to further cook the batch. After the frappé is thoroughly mixed with the batch, spread it out to cool. Fold in edges so as to prevent forming lumps. After batch is sufficiently cold, place it on a kiss machine and cut and wrap in usual manner. It will not be necessary to pull this batch.

SPECIAL FRAPPÉ:		
Nulomoline (invert sugar)	50	lb
Corn syrup	50	lb
Albumen (dissolved in $5^1/_2$ lb water)	3	lb
Gelatin (low grade) (mixed with $2^1/_2$ lb water)	1	lb

Procedure:

Boil the corn syrup to 250 F. Add **Nulomoline**, mix until melted, then turn off heat. Place the batch into a marshmallow beater, add the gelatine-water combination. Start beater in motion, then gradually add the dissolved albumen and beat until light.

Source: Crompton & Knowles

Peanut Butter Kisses

			% by Wt.
Molasses	2	qt	46.00
Butter	4	oz	2.00
Corn syrup	4	lb	30.00
Salt			to suit
Starch			to suit
Peanut butter	3	lb	22.00
			Total 100.00

Procedure:

Put molasses and corn syrup in copper kettle over good fire and stir until well mixed; boil to "slight crack," or about 260 F. Stir in the butter and allow to come to boil again, and pour out on slab. Sprinkle over the batch about a tablespoonful salt and fold in edges as it cools.

Meanwhile, heat peanut butter in a clean kettle and as it gets thin, stir in sufficient starch to make it like dough; if butter is not salty enough, add salt at this time.

Pull molasses batch over hook until done, spread out in oblong shape on slab, putting peanut butter along the center, and folding the batch over it, close ends and seams. Spin out and cut with shears or kiss cutter. As outside will tend to draw away from the center, leaving butter exposed, it will be necessary to mix latter pretty stiff, as above, so it will not run out. Corn syrup will not mix readily with the butter, hence the starch, or powdered sugar, is recommended.

Source: Savage Brothers Co.

Preserved Fruit Jelly Cakes and Bars

			% by Wt.
Sugar	50	lb	28.29
Corn syrup	25	lb	14.14
Nulomoline (invert sugar)	25	lb	14.14
Agar agar (Japanese gelatine)	1.75	lb	0.99
Water	75	lb	42.44
Assorted preserved fruits			to suit
			Total 100.00

Procedure:

First, cover a slab, table top, or wooden trays with waxed paper and place the metal bars in position. Distribute maximum amounts of preserved fruits over the waxed paper, then proceed to prepare the jelly batch as follows:

First soak the agar agar in the water for at least 1 h (preferably overnight). Heat the agar agar and water until it is dissolved then add the sugar and **Nulomoline** and boil the batch to 222–224 F (sea level). Turn off the heat, add the corn syrup and mix well, then add lemon, orange, or other flavors and pour the jelly over the preserved fruit and when it has set it may be cut into bars.

During humid weather it is desirable to sprinkle XXX sugar over the jelly batch.

The preserved fruits may also be placed into large paper cups and the jelly poured directly over the fruit.

Wrap in transparent cellulose.

Source: Crompton & Knowles

Starch Jellies

(Open-Kettle Method)

	% by Wt.
Nusweet "E" corn syrup (43 Bé)	24.00
Sucrose	24.00
Clinjel 260-B corn starch	6.00
Water	46.00
Flavor	to suit
Color	to suit
Citric acid	to suit
	Total 100.00

Procedure:

Place sucrose, corn syrup, and 1520 g water in steam kettle. Apply steam at 40 psi. Slurry corn starch in 400 g water. Add slurry to *boiling* contents of kettle. Maintain boiling. When 73% solids are obtained (approximately 15 min) add acid, color, and flavor: To 99.2% of starch jelly, add 0.1% flavor, 0.5% color, and 0.2% citric acid.

Cast in starch molds. Dry for approximately 48 h in drying cabinet. Retain cabinet temperature at 110–120 F with low humidity until 80% solids are obtained. Remove from molds. Dust by agitating in wide-mesh sieve under steam jet. Dump into granulated sucrose-containing jar. Remove excess sucrose by agitating in sieve.

Source: ADM

Black Walnut Taffy

			% by Wt.
New Orleans molasses	2	qt	52.34
Bicarbonate soda	1	tbs	0.28
Corn syrup	5	lb	45.12
Butter	4	oz	2.26
Black walnut meats			to suit
			Total 100.00

Procedure:

Cook molasses and corn syrup to 256 F; remove from fire, and stir in butter (best grade), and then the soda, and a few nut meats. Pour out on oiled slab, and work in more nuts, until enough have been added. Now make batch into a triangular loaf, and cut off in slices, wrapping in waxed paper.

Source: Savage Brothers Co.

Coconut Taffy

		% by Wt.
Sugar	4 lb	68.08
Water	1 pt	17.02
Coconut	1 lb	12.77
Cream of tartar	0.5 tsp	2.13
		Total 100.00

Procedure:

Mix sugar, cream of tartar and water; cook to the hard crack. Add one fresh coconut and stir until the batch reaches the "soft crack." Pour out on the greased slab and when cool enough to handle pull until white on the hook.

Source: Savage Brothers Co.

Fine White Molasses Taffy

		% by Wt.
Sugar	4 lb	24.25
Golden syrup	2 qt	36.36
Molasses	4 pt	36.36
Vinegar	0.5 pt	3.03
Extract of lemon (to flavor)		to suit
		Total 100.00

Procedure:

Cook granulated sugar and golden syrup together; boil until thick when dropped in water, then add molasses and vinegar. Boil to the crack, remove from fire and stir in quickly a tablespoonful (scant) of soda, flavor with extract of lemon and pour on the slab. Pull over the hook until white—then place on table, pull out into string and cut to suit your fancy.

Source: Savage Brothers Co.

Molasses Sponge Chewing Taffy

Brown sugar	10	lb
Molasses	0.5	gal
Cream of tartar	1	tsp
Milk	1	qt
Hard butter	2	oz

Procedure:

Place in a copper kettle, brown sugar, molasses, and a little water; set on the fire and stir until all the sugar is dissolved. Add a good teaspoonful cream of tartar and allow it to come to a boil, adding gradually 1 qt of good milk and 2 oz hard butter. Cook to 270 F.

Pour on greased slab, and pull over hook until light and spongy, flavoring while on the hook. When done, spread the batch and cut into squares, wrapping in waxed paper.

Source: Savage Brothers Co.

Molasses Taffy

(Wrapped in Cellophane or Waxed Paper)

			% by Wt.
Corn syrup	25	lb	42.29
Molasses (**Grandma's**)	15	lb	25.37
Nulomoline (invert sugar)	5	lb	8.46
Granulated sguar	10	lb	16.91
Salted dairy butter	2	lb	3.38
Hardened coconut oil (92–100 F m.p.)	2	lb	3.38
Salt	2	oz	0.21
			Total 100.00

Procedure:

Place all ingredients together into a kettle and cook to approximately 258–260 F. Then pour the batch out to cool, and when sufficiently firm, pull well on the hook, adding vanilla, rum, or lemon flavor.

Spin, cut, and wrap.

Source: Crompton & Knowles

Molasses Taffy Bars, Cuts, Strips

			% by Wt.
Molasses (**Grandma's**)	10	lb	31.68
Granulated sugar	5	lb	15.84
Nulomoline (invert sugar)	2	lb	6.34
Corn syrup	7	lb	22.18
Dairy butter	2.5	lb	7.92
Frappé No. 3 (q.v.)	5	lb	15.84
Powdered salt	1	oz	0.20
		Total	100.00

Procedure:

Boil the molasses, sugar, **Nulomoline**, corn syrup, and butter to 265–270 F (sea level) then add frappé and salt and mix thoroughly. Spread batch out to set, then later cut into squares, oblongs, strips, or kisses; wrap in cellophane or waxed paper.

FRAPPÉ NO. 3:

Corn syrup	50	lb
Nulomoline (invert sugar)	50	lb
Albumen—dissolved in	3	lb
Water	6	lb

Procedure:

Boil the corn syrup to 245 F. Turn off the heat and add **Nulomoline** and stir until melted. Place the batch into a marshmallow beater; start the beater, then gradually add the albumen solution and beat until light.

Source: Crompton & Knowles

Plain Molasses Taffy

Molasses	2	qt
Baking soda	1	tsp

Procedure:

Put good molasses into a copper kettle and set over a slow fire; boil for about half an hour, stirring and watching to prevent boiling too high. When it begins to thicken add a teaspoonful of dry sifted baking soda; test in ice water and when brittle pour out on a greased slab.

Source: Savage Brothers Co.

Plain White Taffy

Take 6 lb of the best double refined white sugar, to which add 3 pt water. Stir until sugar is dissolved over the fire, add 1 tsp cream of tartar and boil to "the ball." Then add $^{3}/_{4}$ lb butter cut in small pieces, stir with the paddle until melted and thoroughly mixed with the batch. Now add a few drops lemon oil or a tablespoonful vanilla, and boil to the "crack."

Pour out on a greased slab, and cut up when cold, or it may be pulled on the hook until perfectly white, and then drawn out into bars and cut to suit your fancy; or it can be twisted into a mass, and when cold, it can be broken with a hammer.

Glucose (corn syrup) can be added to this recipe, about 2 lb to the above quantity of sugar, omitting cream of tartar.

Source: Savage Brothers Co.

Vanilla Taffy

			% by Wt.
Sugar	6	lb	82.76
Water	1	pt	13.80
Cream of tartar	1	tsp	3.44
		Total	100.00

Procedure:

Use sugar, water, and cream tartar (dissolved in water). Cook until when dropped in water it will crack between the fingers, but if held a moment, forms into a hard ball again (this is called "soft crack"); pour on to greased slab and when cool enough to handle pull over the hook until perfectly white, adding vanilla to flavor when it is on the hook.

Now on the table pull out into strips or flat pieces about 4–5 in. wide and cut in lengths to suit. If very brittle taffy is wanted, cook until crisp when dropped in cold water, and just too hard to form a ball; this is called "the crack."

Source: Savage Brothers Co.

Chocolate Taffy

Use same formula as for vanilla taffy except when poured on the slab, add at once 6 oz melted chocolate and work in with the spatula before pulling.

Source: Savage Brothers Co.

Peanut Taffy

Take 2 lb brown sugar and 2 lb molasses and boil to "the crack," adding $^1/_2$ lb butter cut into small pieces when about two-thirds done. When done, pour on greased marble and scatter roasted peanuts over the surface. When cool, it may be worked into pieces or bars.

Source: Savage Brothers Co.

Salt Water Taffy

No. 1

			% by Wt.
Granulated sugar	10	lb	44.49
Corn syrup	10	lb	44.49
Water	1	qt	8.89
Salt	1	oz	
Butter	6	oz	1.67
Gelatin	2	oz	0.56
			Total 100.00

Procedure:

Melt sugar and corn syrup in 1 qt water. Add 1 oz fine salt and boil to 240 F. Add 6 oz sweet butter and cook to 250 F. Add 2 oz gelatine (dissolved in hot water bath) and cook to 252 F.

Pull over the hook for 15 min, adding flavor and color while pulling.

Source: Savage Brothers Co.

No. 2

	% by Wt.
Clinton Regular corn syrup (43 Bé)	47.50
Sucrose (granulated)	31.60
Water	6.30
Plastic coconut butter (96–100 F m.p.)	3.20
Emulsifier	to suit
Unsweetened evaporated milk	4.00
Salt	4.70
Frappé	2.70
Flavors	to suit
	Total 100.00

Procedure:

Place sucrose, corn syrup, and water in kettle; heat with stirring until sucrose is thoroughly dissolved. Add milk, coconut butter, and emulsifier slowly while mixing. Cook to 256–260 F depending on season of year and local climatic conditions. Add frappé and salt and mix well.

Pour onto oiled cooking slab. Fold batch several times while cooking. Place on pulling machine or pull by hand. Add flavors while pulling. Pull for approximately 3 min. Stripe candy as desired. Cut and wrap pieces on kiss machine, or cut into sticks of desired length and wrap.

Source: ADM

Sea Shore Chewing Taffy

			% by Wt.
Granulated sugar	10	lb	36.94
Water	2	lb	7.39
Corn syrup	10	lb	36.93
Nulomoline (invert sugar)	4	lb	14.77

Hardened edible oil (86–96 F m.p.)	1	lb	3.69
Salt	1	oz	0.28
Color and flavor			to suit
			Total 100.00

Procedure:

Place the sugar, water, corn syrup, and **Nulomoline** into a kettle and heat to the boiling point. Wash down all sugar crystals that adhere to the kettle, add the edible oil and cook the batch to a medium-firm ball (254–262 F depending on prevailing outside temperature).

Pour the batch onto an oiled cooling slab; add the salt. When partly cool, fold in the edges. When sufficiently cool, pull well on the hook, adding color and flavor as desired. Spin, cut, and wrap in waxed paper or cellophane.

Source: Crompton & Knowles

Taffy Apples

Cook sugar and corn syrup, in equal parts, to 248 F; have ready some clean, firm apples with wood skewers driven through them. Take batch from fire, set it in a hot water bath, adding the juice of a lemon and dip the apples quickly, turning them over until well coated, and lay on waxed paper to harden. If troubled with stickiness, cook to 255 or even 260 F but no higher.

Source: Savage Brothers Co.

Yale's New England Peanut Taffy

			% by Wt.
Sugar	9	lb	33.33
Corn syrup	3	lb	11.11
Water	3	pt	11.11
New Orleans molasses	1	qt	11.11
Butter	8	oz	1.85
Roasted peanuts	8	lb	29.64
Sodium bicarbonate	2	tsp	1.85
			Total 100.00

Procedure:

Using a medium-sized kettle, cook sugar, glucose (corn syrup), and water to a "ball;" then add New Orleans molasses, butter, and hot peanuts (previously roasted and thoroughly cleaned) and stir constantly until the thermometer reaches 268 F. Now remove from the fire and stir in 2 tsp sodium bicarbonate, and when the batch begins to foam, pour on greased slab very thin—spread out quickly with palette knife and pull still thinner, and then cut in bars about 2 × 5 in.

Source: Savage Brothers Co.

Butterscotch Toffee Brittle

		% by Wt.
Fine granulated sugar	9 lb	46.01
Nulomoline (invert sugar)	1 lb	5.11
Dairy butter (unsalted)	8 lb	40.89
Whole almonds	1.5 lb	7.67
Salt	1 oz	0.32
Vanilla flavor		to suit
		Total 100.00

Procedure:

Place the butter into a clean copper kettle, heat it slowly until melted, add sugar; stir well then add **Nulomoline.** Allow batch to cook on a moderate fire to boiling point, then wash down sides of kettle, using smallest amount of water possible. Continue to stir and boil the batch to approximately 260 F.

Now add the almonds which should be unroasted. Continue to stir batch until it becomes "livery" in consistency; that is, similar to very heavy paint or a stiff jelly. This is the critical point and temperature will be approximately 290 F. The consistency varies considerably with the quality of the butter used.

After batch has been cooked to consistency described above, add salt, flavor, and spread immediately on an oiled slab with a palette knife; level it off and evenly distribute nuts throughout batch. Cut batch into pieces with roller knife and break paper when cold. Then coat in a combination of milk chocolate mixed with ground roasted almonds, Brazils, etc.

Source: Crompton & Knowles

English Wrapped Toffee

			% by Wt.
Corn syrup	38	lb	27.47
Nulomoline (invert sugar)	10	lb	7.23
Dairy butter (salted)	5	lb	3.62
Coconut butter (86 F m.p.)	10	lb	7.23
Condensed milk (whole 40% sugar)	40	lb	28.91
No. 7 sugar	15	lb	10.85
Fine granulated sugar	20	lb	14.46
Salt	5	oz	0.23
Combination of clove oil, sweet almond oil, and vanilla flavor			to suit
			Total 100.00

Procedure:

Place into a caramel cooking kettle the **Nulomoline**, corn syrup, butter, coconut butter, and condensed milk. Give the steam valve a couple of turns and allow the batch to mix until the fats are dissolved and mixed with other ingredients. Then add the brown and white sugars, turn the steam on full and cook the batch as rapidly as possible to a medium-firm ball.

This toffee can be cooked quite successfully using as much as 110 lb of steam pressure. After the batch has been cooked to desired consistency, turn off heat, add the salt and flavor and mix thoroughly. Spread the batch out at once to cool, then gradually fold it together so as to secure an even temperature throughout the batch. It is then ready to cut and wrap on the kiss wrapping machine.

This toffee can also be cut into squares and wrapped in waxed paper, twisting the ends of the waxed paper instead of folding it.

NOTE:

For each 550 ft above sea level, reduce the boiling temperature by 1 F.

Source: Crompton & Knowles

Rum-Butter Toffee

(Wrapped)

			% by Wt.
Condensed whole milk	40	lb	30.71
Corn syrup	30	lb	23.02

Nulomoline (invert sugar)	10	lb	7.67
Molasses (**Grandma's**)	5	lb	3.83
Fine granulated sugar	30	lb	23.02
Hard coconut butter #90	15	lb	11.51
Salt	5	oz	0.24
Rum and butter flavors			to suit
			Total 100.00

Procedure:

Place condensed milk, corn syrup, **Nulomoline**, molasses, and hard coconut butter into a caramel cooking kettle, equipped with double-action mixers. Turn the heat on partly; mix batch until the fat is completely dissolved, then add granulated sugar, mix well, turn steam on full and boil the batch as rapidly as possible to a medium firm ball. Turn off heat, add salt, then rum–butter flavors and mix thoroughly.

Spread the batch out to cool, then cut into pieces of the desired size and wrap.

Source: Crompton & Knowles

Toffee Nut Squares

			% by Wt.
Corn syrup	20	lb	21.05
Condensed milk (whole or skimmed)	30	lb	31.58
Nulomoline (invert sugar)	10	lb	10.53
Dairy butter	5	lb	5.26
Sugar	25	lb	26.32
Hard coconut butter	5	lb	5.26
Flavor combination: vanilla, maple, and cinnamon			to suit
			Total 100.00

Procedure:

Place the corn syrup, condensed milk, **Nulomoline**, dairy butter, and coconut butter into a kettle, turn heat on partly and mix batch until fats are dissolved. Then add the sugar (no water) and boil the batch to a firm ball. Turn off the heat, add salt and flavors and mix well.

Spread the batch on an oiled slab and immediately mark the batch into squares or pieces of the desired size. When cold, cut and wrap in waxed paper or transparent cellulose.

Source: Crompton & Knowles

Super-Fine Toffee Butter Crunch

			% by Wt.
Salted dairy butter	5	lb	21.56
Hardened edible oil (76 F m.p.)	7	lb	30.19
Lecithin	0.5	oz	0.13
Granulated sugar	18	lb	34.51
Nulomoline (invert sugar)	1.5	oz	0.40
Molasses (**Grandma's**)	2	lb	8.63
Water	1	lb	4.31
Salt	1	oz	0.27
Vanilla or maple flavor			to suit
			Total 100.00

Procedure:

Melt butter and edible oil, add lecithin and stir until lecithin is dissolved. Add sugar, mix well, then add **Nulomoline**, molasses, water, and continue to stir; cook batch to approximately 300–305 F. Add salt mixed with a little vanilla or maple flavor and mix thoroughly.

Spread the batch thinly on a cool slab. While batch is still slightly warm, grind it into pieces of the desired size, sieving or screening off the various particle sizes.

Apply a little melted edible oil to the candy particles mixing well, then dust slightly with powdered sugar so that the candy particles when they have set will not adhere together.

Source: Crompton & Knowles

Almond Tutti-Frutti Chews

			% by Wt.
Coconut (medium shred)	14	lb	28.82
Granulated sugar	10.5	lb	21.61

Buttered-diced-roasted almonds	5	lb	10.29
Diced mixed glace fruit	3	lb	6.17
Egg whites	10	oz	1.28
Corn syrup solids	15	lb	30.88
Salt	2	oz	0.25
Cocoa butter	4	oz	0.51
Vanilla	1.5	oz	0.19
		Total	100.00

Procedure:

PART 1: Place coconut in mixer. Combine 4 lb of the sugar with $4^1/_2$ lb hot water. Heat until dissolved and blend with coconut about 3 min at low speed.

PART 2: Beat egg whites until stiff. Combine $6^1/_2$ lb of the corn syrup solids with remaining sugar and blend dry. Add 7 lb water and salt and cook to 230 F. Add this syrup one-third at a time to beaten egg whites. Total time for mixing about 8 min. Pour frappé (Part II) over sweetened coconut (Part I) at low speed. Add almonds and fruit and mix.

PART 3: Melt cocoa butter and add to above. Add vanilla, then add remaining corn syrup solids and mix until blended. Remove from mixer and cool to room temperature. Extrude on enrober feed belt with rotary-type extruder or form as desired and coat with chocolate.

Source: California Almond Growers Exchange

Almond Vinegar Cuts

			% by Wt.
Brown sugar	15	lb	15.74
Corn syrup	15	lb	15.74
Dairy salted butter	1	lb	1.05
Hard coconut butter	2	lb	2.10
Manufacturing cream (20% butterfat)	2	gal	17.83
Evaporated milk	2	gal	18.69
Vinegar	2.5	pt	2.62
Select sheller run almonds, roasted and chopped coarsely	15	lb	15.74
Select sheller run almonds, roasted and ground fine	10	lb	10.49
		Total	100.00

Procedure:

Cook sugar, corn syrup, coconut butter, and manufacturing cream in double-action fire mixer. After boil is reached add evaporated milk slowly as not to break boil. Cook to high caramel test. Add vinegar slowly, do not break boil. Add dairy butter. Cook back to firm ball caramel test. Turn off fire and add chopped roasted almonds. Mix in thoroughly. Pour batch on oiled steel slab to desired height. When cool cut in bars or squares and dip in milk chocolate into which ground roasted almonds have been mixed.

Source: California Almond Growers Exchange

Turkish Cuts or Cast Gums

			% by Wt.
Corn syrup	65	lb	30.61
Nulomoline (invert sugar)	10	lb	4.71
Thin boiling starch # 65	12	lb	5.65
Cream of tartar	6	oz	0.18
Water	100	lb	47.08
Granulated sugar	25	lb	11.77
Color and flavor			to suit
			Total 100.00

Procedure:

Place the corn syrup, **Nulomoline**, starch, cream of tartar and water into a kettle, gradually bring to the boiling point and allow to simmer for 5 min, then turn the steam on full and boil the batch to a good string.

Add the granulated sugar and boil the batch slowly until the sugar is dissolved, then add color and flavor and cast the batch into starch or pour it into trays lined with heavy wrapping paper.

In the event the jelly is cast into starch, it should remain in a drying room for 3–4 days at a temperature of approximately 135 F, then remove from the starch, spray with oil and later pack.

If a cut piece is desired, then permit the jelly to remain in the trays for at least 4 days, after which it may be cut into pieces of the desired size and rolled in XXX sugar.

Source: Crompton & Knowles

Turkish Paste

(Cut)

			% by Wt.
Granulated sugar	55	lb	24.87
Nulomoline (invert sugar)	45	lb	20.34
Cold water	110	lb	49.74
Thick boiling starch	11	lb	4.97
Cream of tartar	3	oz	0.08
			Total 100.00

Procedure:

Place 50 lb of the water into a kettle, add the sugar and **Nulomoline** and allow the batch to boil.

Meanwhile, mix the starch and cream of tartar with the balance of the water and gradually add this to the boiling batch and cook continuously using approximately 60 lb of steam pressure. Boil the batch to a heavy sheet.

In order to test this place a small amount of the hot batch between the thumb and forefinger and blow cool air between the fingers and work them back and forth. When all of the paste leaves the forefinger and sticks to the thumb, it may be considered properly cooked.

After the batch has been properly cooked and flavored, pour it into trays lined with wrapping paper or into starch trays which are covered, sides and bottom, with approximately $^1/_4$ in. of dry starch. Sprinkle starch over the surface of the sheets of paste in the starch trays and allow them to set for 2–4 days. Then the sheets are ready to cut. After the paste has been cut into pieces, roll them in a mixture of 90% sugar and 10% dry corn starch and later pack.

Source: Crompton & Knowles

Chapter 4

CHOCOLATE PRODUCTS

Chocolate: Its Source and Preparation

It is useful to look briefly at the sources of cocoa beans and the steps taken to ready the beans for processing.

The cacao trees (*Theobroma cacao*) grow in humid tropical areas located 20 degrees north or south of the equator. These areas extend through South and Central America, the Caribbean, midwestern Africa, nearby islands, to the islands of Java, Samoa, Malaysia, the Philippines, and portions of India.

The seeds or beans are enclosed in a thick-walled pod that grows attached to the large branches and/or to the main trunk of the tree, unlike other fruits that grow along the thin branches.

When the pods reach maturity, they are collected, usually twice a year. The pods are split open and the seeds and the sweet mucilage that surrounds them are scraped into bags or boxes. The containers are carried to fermentation beds where the carbohydrate-rich mucilage is subjected to the action of microorganisms (wild yeast), resulting in alcoholic fermentation. Next the acetic bacteria turn the alcohol into acetic acid (vinegar).

As soon as this dripping mass drains into the ground, the beans are spread into drying beds where the heat of the sun or artificial heat reduces the moisture of the beans to about 8–9%.

A safe total moisture is necessary before the beans are bagged for shipment to avoid the formation of mold that may make them unfit to produce good liquor. If cocoa beans are allowed to become moldy, the off-flavor produced by the mold will be carried by the liquor all through the manufacturing of chocolate and cocoa powder.

The pH of a typical liquor fluctuates between 5.2 to 5.6 with the exception of some unusually low pHs found lately in beans and liquor from Malaysia of only 4.0–4.4.

The cocoa beans are received at the plant in bags weighing about 140–160 lb. They are immediately sent through the cleaning and destoning equipment before being stored in silos or sent directly to the roasters. Roasting is followed by the deshelling units (crackers and fanners) and the stone crushing, hammer mill system or any other crushing system to produce liquor. Since the nibs contain from 50–60% cocoa butter, friction heat upon crushing liquefies the cocoa butter, producing a brown mass normally called chocolate liquor, cocoa mass, or bitter chocolate.

Chocolate Terminology

Chocolate liquor is produced by crushing and refining the roasted cocoa nibs. It contains no less than 50% and as much as 56% cocoa butter. In this form it is called chocolate liquor, cocoa mass, bitter chocolate, baking chocolate, or cooking chocolate.

Chocolate liquor may be referred to by the name of the region from which the cocoa beans come—Caracas, Maracaibo, Arriba, Sanchez, or Bahia—or of the country of origin, as African liquor, Java, Ecuadorina, Brazilian, and so on.

Chocolate liquor is classified as natural or American processed, meaning that no chemical changes have occurred in the liquor, or as dutched processed, meaning that the liquor has been alkalized or chemically treated with potassium carbonate, sodium bicarbonate, or any of the alkaline salts permissible in the process.

Since cocoa beans vary in color from a very light creamy to a very dark brown, the names of the liquors may also represent the types of beans used. Java will be the lightest, and Arriba will be one of the darker. The alkalized liquors will also vary in intensity of color according to pH, roasting, and other treatments.

Chocolate liquor may be used directly as a coating or as an ingredient in industries such as baking, confectionery, ice cream, flavoring, and tobacco. Mainly it is used to manufacture chocolate coatings (milk or dark), cocoa powders, ice cream mixes, cookies or cakes, and other confections.

Dark chocolate coating is the product of a refined blend of liquor, sugar, cocoa butter, and flavor. It could be dark, medium, or light and be sweet or semisweet or bittersweet depending on the amount of liquor used in the formula.

Milk chocolate coating is the product of a refined blend of sugar, milk solids, liquor, cocoa butter, and flavor. Lecithin and salt may be used in the formulation. Either milk or dark coatings can be made with endless variations in the proportion, fineness, and viscosities of their ingredients.

Manufacturing Chocolate Coatings

In the manufacturing of coatings there is a constant enigmatic question: What system do you use to make coating? This old, reliable system is suggested: Blend sugar, milk solids, liquor, and a portion of the cocoa butter to create a semidry paste. Transfer this paste to a five-roll refiner; feed conches with the refined paste and conche such paste for "X" number of hours at "Y" temperature. Correct viscosity as desired and mold, store, or ship liquid. Dark chocolate coating follows a procedure just like milk coating except that milk is not normally an ingredient.

For manufacturing chocolate coatings, there are at least a half-dozen ways to refine chocolate coating masses. Some of the available systems are: the Bramley Mill, the MacIntyre Mill, Ball Mills, Micro-Pulverizing, and finally the rolls refiners. In addition, the manufacture of milk chocolate using milk–sugar–crumb or milk–sugar–liquor–crumb is typical of most European and some American coatings processing.

A frequently asked question is: What makes a chocolate coating high quality and what makes it low quality when both must meet Government standards?

To elucidate this question, some formulations are presented, two for dark coatings and two for milk coatings, as well as some of their physical specifications. A higher price can be justified by comparing #1 and #2.

Basic Chocolate Coating Formulas

#1 Dark Chocolate

	% by Wt.
Sugar	48.00
Chocolate liquor	35.00
Cocoa butter	16.95
Vanilla and/or vanillin	0.05
Lecithin (if desired @ 2 oz/100 lb)	
	Total 100.00

Fineness — 0.0005 in.
Conche — 32 h
(Conching for 12 h at 120 F, 12 h at 160 F, and 8 h at 120 F)
Total fat — 35.0%

#2 Dark Sweet Chocolate

	% by Wt.
Sugar	60.50
Chocolate liquor	15.00
Cocoa butter	24.20
Vanillin	0.05
Lecithin	0.25
	Total 100.00

Fineness — 0.0012 in.
Conche — 8 h
(Conching for 6 h at 140 F and 2 h cooling to 120 F)
Total fat — 32.0%

Fineness is given by the micrometer caliper or by microscope projections. Fat analyses are measured by petroleum ether extraction or by Halowax oil, refractive index.

#1 Milk Chocolate

	% by Wt.
Sugar	40.95
Cocoa butter	22.00
Whole milk (spray-dried)	20.00
Chocolate liquor	17.00
Vanillin	0.05
	Total 100.00

Fineness — 0.004–0.005 in.
Conche — 48 h
(Conching for 12 h at 120 F, 12 h at 140 F, 8 h at 160 F, 10 h at 140 F, 6 h at 110 F.)
Total fat — 36.5%

#2 Sweet Milk Chocolate

	% by Wt.
Sugar	54.00
Cocoa butter	22.70
Whole milk	13.00
Chocolate liquor	10.00
Lecithin	0.25
Vanillin	0.05
	Total 100.00

Fineness — 0.0012 in.
Conche — 8 h
(Conching for 6 h at 140 F and 2 h to 110 F.)
Total fat — 32.0%

The mixed mass is transferred to refiners, which reduce the particle size of the ingredients to a predetermined size.

The refined mass goes into conches where the heat and agitation produce a flavor that has also been predetermined in each case.

After conching, the mass goes into tanks for storage or any other needed correction such as viscosity or color, thus making the finished product ready for molding or shipping in liquid tank wagons.

The chocolate liquor will probably be a blend of liquors in order to obtain a more pleasant flavor. It may be a natural or dutch blend.

Lecithin is an optional ingredient in chocolate; it is welcomed by some chocolate users and rejected by others. That applies to dark as well as to milk chocolate coatings.

Chocolate Drops

One other chocolate item that has become extremely popular in the baking and ice cream industries is chocolate drops (bits, chips, or whatever). These drops are made just as are coatings, except that the mixture is deposited on cold steel belts to a predetermined size or count, which, in turn, will give a certain number of drops per pound. The number or counts normally made for various users may be 950, 1000, 2000, 4000, 5000, 6500, or 10,000 drops per pound.

Compound Drops. Compound drops are manufactured just as confectioners coating and deposited like the chocolate drops. Those containing lauric fat are made with low-fat cocoa powder, and those made with nonlauric fat may contain chocolate liquor only, a liquor–cocoa blend, or cocoa powder alone.

Pastel Drops or Pastel Confetti or Coatings. These are made with vegetable fats, milk solids, and sugar, to which a certified color may be added, and come in white, green, pink, yellow, and other colors. In addition, drops, confetti, and coatings are made with peanut butter, butterscotch, coconut, lemon, cinnamon, and so on.

Fats in Compound Coatings or Drops

The fats used in compounds are classified as lauric and nonlauric according to the presence of absence of lauric fatty acid in the fatty molecule. If the molecule contains lauric fatty acid, the fat is called lauric fat. If not, the fat is called nonlauric fat.

Typical nonlauric fats are: cocoa butter, soyabean, cottonseed, palm, peanut, shea butter, and ellipi butter. Typical lauric fats are: coconut, palm kernel, and babassu.

The methods by which these hard butters (lauric and nonlauric) are manufactured vary significantly within the various processors.

The methods can be described as: fractionated, partially or totally hydrogenated, winterized, and interesterified.

Depending on the method used to produce the hard butters and the blend of oils, the hard butter may have a pleasant texture, quick melting in the mouth, or may be chewable or waxy and not so pleasant.

Typical Compound Formulas and Their Manufacture

For the sake of demonstration, the following are typical compound formulas:

Nonlauric (NL) Types

Milk Chocolate

	% by Wt.
Sugar	50.70
Hard butter	30.00
Nonfat dry milk	12.00
Chocolate liquor	5.00

Cocoa powder	2.00
Lecithin	0.25
Vanillin	0.05
	Total 100.00

Total fat — 33.05
Fineness — 0.007 in.
Viscosity — NCA @ 110

Dark Chocolate

	% by Wt.
Sugar	59.00
Hard butter	28.70
Chocolate liquor	12.00
Lecithin	0.25
Vanillin	0.05
	Total 100.00

Total fat — 35.0%
Fineness — 0.007 in.
Viscosity — NCA @ 110

Lauric (L) Types

Milk Chocolate

	% by Wt.
Sugar	47.70
Hard butter	30.00
Nonfat milk	12.00
Cocoa powder (L-F*)	10.00
Lecithin	0.25
Vanillin	0.05
	Total 100.00

Total fat — 31.0%
Fineness — 0.007 in.
Viscosity — NCA @ 120

*L-F = lowfat cocoa

Dark Chocolate

	% by Wt.
Sugar	56.00
Hard butter	30.70
Cocoa powder (L-F-D*)	12.00
Emulsifier** (other than lecithin)	1.00
Lecithin	0.25
Vanillin	0.05
	Total 100.00

Total fat — 33.0%
Fineness — 0.0007 in.
Viscosity — NCA @ 110

Tempering and Usage of Coatings

The tempering of chocolate coatings is a must in order to obtain acceptable appearance and an extensive shelf life. Tempering can be done by machine, by an experienced hand operator, or by seeding the coating. The fat crystal formation must be of a size that will give the solidified coating a durable hard consistency.

A tempering unit does the tempering automatically once it is set.

A manual tempering operation, in general, will melt the coating up to about 120 F in order to destroy the fat crystals present, then slowly cool it down to about 79–80 F for milk and 80–82 F for dark, maintain a slow agitation for 20–30 min, and start raising the temperature slowly to 86–88 F for milk and 88–92 F for dark. At this point it can be used.

The seeding system of tempering is done by adding fine shavings made from a well-tempered bar to a coating at 94–96 F for milk and 96–98 F for dark; after the coating is allowed to seed, and the temperature is reduced about 6–8°, the coating is put to use.

One other method sometimes used is to slowly melt a well-tempered coating, making sure that the temperature *never* goes over 100 F. When the whole mass is melted, it is cooled down and used.

* L-F-D = lowfat dutch cocoa

** Emulsifiers may be Polysorbate 60 and sorbitan monostearate or other permitted by FDA.

Tempering Compound Coatings is somewhat easier than tempering chocolate. The nonlauric fat coatings do not require tempering (as recommended by the fat manufacturer); they can be used at 104–109 F.

For the lauric fats, since there are so many melting point variables, the simplest method is as follows: Melt coating to 15–20° above the melting point. Slowly cool it down to a mush consistency and maintain for about 15–20 min. Take the temperature of the mass, and raise the temperature to 10–15 F from that point and use the coating.

Some lauric fats are previously seeded by the supplier or by the coating manufacturer, making the coating self-tempering as long as the enrobing temperature is within 10° of the fat melting point.

Enrobing

Many factors should be considered in producing perfectly enrobed items.

The tempering of the coating is necessary, but in the hardening process, the freshly coated piece must be exposed to correct temperatures and relative humidity before a shiny, hard, and durable coating is obtained for a long shelf life.

The viscosity of the coating to be used in enrobing by machine varies significantly based on the amount of coating the user wants to have on the centers. Some require a viscosity of 50° NCA* so that the amount of coating on the center will be minimal; others require 110–120° NCA for an extensive milage but still have enough coating on the center so the chocolate will be noticeable in the eating. Finally, there is the manufacturer who sells as much chocolate as center and requires viscosities in excess of 165° NCA.

It must be mentioned that, as part of enrobing, pre-bottoming is very much in use to assure a firm base, especially for soft or creamy centers.

The following are ideal conditions of a cooling tunnel: Length: No less than 95 ft long, temperature at entrance of tunnel 60 F, about 45–50 F in center, and 60–65 F at exit. Depending on size and thickness of coating, travel time should be 10–14 min. Relative humidity 50%. Store candy at 65–70 F for 24–48 h to expel heat of crystallization (latent heat).

If such equipment cannot be afforded, it is possible to adapt the situation to the best available conditions, such as hardening and recovering rooms with humidity control.

*NCA = National Confectioners Association. These are given in "degrees MacMichael." 50 is thinner than 100.

Milk and Dark Chocolate Usage

Most chocolate coatings, both milk and dark, are used to enrobe centers. This is done either by machine or by hand dipping.

The centers are numerous; they vary from creams and nougats to solid nuts and pretzels. When coating is used for bars, the centers also offer an extensive variety. Many combinations of nougat, caramel, and peanut butter and, more recently, baked items have been used as centers or as part of a cookie, waffle, crunchy biscuit, or some other baked product.

A great deal of chocolate is used in bars, either solid or blended with peanuts, almonds, or hazelnuts or in "bark" which is just coating mixed with whole nuts, normally almonds, then spread on a cold table for hardening and sold in broken irregular pieces.

All of these items may be reproduced with a good quality confectionery coating and/or with pastels or other flavor coatings.

Ice Cream Chocolate Coatings

An item dating back many years is the ice cream coating, originally made with chocolate coating diluted in cocoa butter to an extent of about 60% total fat. As time passed, the added cocoa butter was replaced with an oil that had a lower melting point. Coconut and palm kernel oils were preferred to any other for their fast hardening time. They are still the most commonly used for that purpose, even though quite frequently a blend of coconut and soya oils is used.

Typical formulas of ice cream coatings are as follows:

#1 Ice Cream Coating

	% by Wt.
Chocolate liquor	15.00
Sugar	34.60
Lecithin	0.35
Vanillin	0.05
Coconut oil (76 F)	50.00
	Total 100.00

Fineness — 0.0008 in.
Total fat — 58.0%

#2 Ice Cream Coating

	% by Wt.
Cocoa powder	8.00
Sugar	34.60
Lecithin	0.35
Vanillin	0.05
Coconut oil (76 F)	57.00
	Total 100.00

Fineness — 0.0008 in.
Total fat — 58.0%

There is an option to reduce the total fat in these formulas to about 35–40% and use it to spoon over ice cream served in cups or as a dip for cones. For such uses as well as for bar coatings, the coating must be maintained warm at about 95 F for small use to 105 F for machine dipping.

The ice cream coating formulas #1 and #2 with the reduced total fat could well be adapted to a melt-away type of refrigerated bar. The bar will be more attractive when hazelnut paste is incorporated into the mass.

Spreads

The following formula can be used for chocolate, cocoa, peanut butter, and hazelnut butter to spread on bread, crackers, or cakes.

	#1	#2	#3	#4
Chocolate liquor	23.8	—	—	—
Cocoa powder	—	13.8	7.3	7.3
Sugar	55.5	55.5	45.0	45.0
Butter (peanut or hazelnut)	—	—	27.0	27.0
Peanut oil	20.0	30.0	20.0	20.0
Salt (optional)	0.1	0.1	0.2	0.2
Mono- and diglycerides (optional)	0.5	0.5	0.5	0.5
Vanilla or vanillin (optional)	0.1	0.1	—	—
Totals	100.0	100.0	100.0	100.0

Fineness — 0.0010 in.
Mix to homogenize; bottle or carton and refrigerate.

Chocolate is still the "flavor of preference," even though vanilla is, in certain areas, still classified as number one. In any case, the acceptability of chocolate is so great that the pride of producing a good-tasting, eye-appealing, and enjoyable-to-eat product must be maintained.

Determining Viscosity of Chocolate and Compound Coatings

APPARATUS:

MacMichael Viscosimter
#26 wire
2 cm bob
Oven capable of maintaining 120 F
Aluminum or copper cup
Thermometer (F)
Stirring rod

PROCEDURE:

1. Equilibrate viscosity cup and bob in oven at 120 F before use.
2. Turn on heat switch of viscosimeter.
3. Both chocolate and coating must reach a temperature of 118 F before removing from oven.
4. Mix well with stirring rod.
5. Cool chocolate to 102 F and coating to 112 F.
6. Place cup and wire in position on viscosimeter.
7. Pour chocolate or coating into cup filling to second line from top on bob.
8. Read chocolate at 100 F and coating at 110 F.
9. Read and record the viscosity indicated on the scale to the nearest whole number.

NOTES:

1. Speed of cup rotation should be set at 15 rpm and checked frequently.
2. Inaccurate results can be obtained from using a bent wire.
3. Keep instrument level to prevent bob shaft from binding on collar.
4. Avoid getting water into chocolate or equipment.

Almond Paste Chocolate Centers

			% by Wt.
Nulomoline (invert sugar)	15	lb	14.29
Corn syrup	30	lb	28.57
Natural almond paste (California Almond Growers Exchange)	30	lb	28.57
Liquor chocolate	30	lb	28.57
Salt and vanilla flavor			to suit
			Total 100.00

Procedure:

Heat the **Nulomoline** until melted. Add the corn syrup. Mix well. Add the salt, flavor, and almond paste. Mix until smooth.

Add the liquor chocolate which has been melted and cooled to 90 F. Mix thoroughly and spread out to set.

Centers may be formed by machine or hand., coated with chocolate, icing, or rolled in colored sprinkles. Sheets may be cut into squares, oblongs, etc., and packed in paper cups.

Source: Crompton & Knowles

Almond Truffles

			% by Wt.
Milk chocolate	10	lb	16.60
Dark chocolate	10	lb	16.60
Vegetable fat (92 F)	20	lb	33.20
Roasted small steel-cut almonds	20	lb	33.20
Lecithin	1	oz	0.10
Fine flour salt	3	oz	0.30
			Total 100.00

Procedure:

Melt chocolate, vegetable fat, salt, and lecithin to 110 F. Cool to 75 F and place in a planetary-type beater with wire whip and beat for approximately 6 min or until light. Mix in the almonds and spread on paper-lined slab. Before it is completely set, cut with hot, wet roller knife in about $^3/_4$ in. squares. Let set until firm and coat with milk or dark chocolate.

Source: California Almond Growers Exchange

Chocolate Clusters

			% by Wt.
Milk chocolate	1	lb	68.58
Bitsyn® or pecan pieces	11	oz	31.42
			Total 100.00

Procedure:

Melt and temper chocolate, mix with **Bitsyn®** and form clusters.

Source: Pillsbury Co.

Chocolate Coconut Clusters

			% by Wt.
Milk chocolate	56	lb	56.0
Toasted shredded coconut	44	lb	44.0
			Total 100.00

Procedure:

Melt chocolate and add toasted, shredded coconut. Stir gently until well mixed. Deposit small clusters and enrobe in milk chocolate as desired.

Source: Kraft General Foods Corp.

Chocolate Coconut Truffles

		% by Wt.
Dark sweet chocolate coating	38.50 lb	41.74
Creamed Coconut	12.75 lb	13.82
Sweetened condensed milk	23.00 lb	24.93
Invert sugar	12.75 lb	13.82
Orange oil or flavors		to suit
Tender Fresh Medium Coconut	5.25 lb	5.69
Walnuts, pecans, or brazil nuts		to suit
		Total 100.00

Procedure:

Melt chocolate and **Creamed Coconut** together and place in mixer. Add sweetened condensed milk, invert sugar, and flavor. Mix well.

Add **Tender Fresh Medium Coconut** and chipped nuts to mixture. Pour on slab to set. When firm, cut into squares.

Source: Kraft General Foods Corp.

Chocolate Patties

		% by Wt.
Sugar	10 lb	62.50
Corn syrup	2.5 lb	13.50
Water	2 lb	12.0
Chocolate liquor	1 qt	12.0
Vanilla bean seed or cinnamon oil		to suit
		Total 100.00

Procedure:

Cook sugar, glucose (corn syrup) with water to 236 F and before the batch is worked, add 2 lb good chocolate liquor and the seeds of one vanilla bean or a drop or two of cinnamon oil. Great care should be exercised, for these wafers will spot more easily than any other color or flavor.

Do not heat the cream more than is absolutely necessary.

Source: Savage Brothers Co.

Chocolate Crunch Shell

		% by Wt.
Milk chocolate	80 lb	80.0
Toasted shredded coconut	20 lb	20.0
		Total 100.00

Procedure:

Melt chocolate and add toasted shredded coconut. Mix well.

Source: Kraft General Foods Corp.

Currant/Crunch Molded Bar

			% by Wt.
Milk chocolate	2	lb	76.20
Zante currants	5	oz	11.90
Bitsyn®	5	oz	11.90
			Total 100.00

Procedure:

Mix tempered chocolate, currants, and **Bitsyn®** and mold.

Source: Pillsbury Co.

Dark Chocolate

	% by Wt.
Sugar (4X)	44.00
Chocolate liquor (53% CB, Accra-Caracas 50-50)	42.00
Cocoa butter	13.30
Vanilla (powder or paste)	0.50
Lecithin	0.20
	Total 100.00

Total fat @ 35%
Fineness — 0.0005 in.
Viscosity @ 120 ± 5 Mac. at 100 F

Procedure:

Conche: 48 h @ 180 F.
Add vanilla at the end of conching when temperature is 140 F.

Source: Wilbur Chocolate Co.

Milk Chocolate

No. 1

	% by Wt.
Sugar	45.45
Chocolate liquor (Accra-Trinidad 75–25*)	17.00
Whole spray powder milk**	20.00
Cocoa butter (prime press)	17.30
Lecithin	0.20
Vanillin	0.05
	Total 100.00

Total fat @ 32%
Fineness — 0.0005 in.
Viscosity @ 120 ± 5

Procedure:
Conche: 48 h @ 160 F.
Add vanilla in conche.

Source: Wilbur Chocolate Co.

No. 2

	% by Wt.
Chocolate liquor (54% fat)	12.00
Whole milk powder (28% fat)	14.00
Anhydrous dextrose (**Chocolose 2403**)	15.30
Sucrose	35.70
Lecithin	0.30
Ethyl vanillin	0.10
Salt	0.05
Added cocoa butter	22.55
	Total 100.00

* You may substitute Bahia liquor
** You may substitute 1.5% with cultured milk

Procedure:

Ingredients are mixed together in preliminary ingredient mix, followed by roll refining and conching.

Source: CPC International

Sweet Chocolate

	% by Wt.
Chocolate liquor (54% fat)	35.00
Sucrose	35.00
Anhydrous dextrose (**Chocolose 2403**)	15.00
Lecithin	0.30
Ethyl vanillin	0.10
Salt	0.05
Added cocoa butter	14.55
	Total 100.00

Procedure:

Ingredients are mixed together in preliminary ingredient mix, followed by roll refining and conching.

Source: CPC International

Plastic Chocolate Centers

			% by Wt.
Chocolate liquor	40	lb	40.00
Nulomoline (invert sugar)	50	lb	50.00
Frappé No. 1 (q.v.)	10	lb	10.00
Peppermint oil			to suit
Powdered salt			to suit
			Total 100.00

Procedure:

Heat the chocolate liquor until melted, then allow it to cool to approximately 90 F.

Heat the **Nulomoline** until clear, allow it to cool to approximately 80–85

F, then add to the chocolate liquor. Mix until smooth. Add the peppermint oil, salt, and Frappé No. 1 and mix thoroughly.

Spread on waxed paper to set, after which it may be formed into centers or spread between a batch of cut marshmallow. The centers may be coated with chocolate, or rolled in ground nuts or chocolate and colored sprinkles.

FRAPPÉ NO. 1:

Corn syrup	50 lb
Invert sugar	50 lb
Albumen	1 lb
Water	2 lb

Procedure:

Cook corn syrup to 245 F (sea level), shut off the heat and add invert sugar and stir until it is melted. Place the batch into a marshmallow beater, start the beater, then gradually add albumen which has been previously dissolved in water and beat the batch until light.

Source: Crompton & Knowles

Turtle-Type Clusters

		% by Wt.
Corn syrup (regular)	420 g	30.51
Whole sweetened condensed milk	880 g	63.93
Satina fat	75 g	5.45
Salt	1.5 g	0.11
		Total 100.00

Procedure:

Cook over very low fire with continuous agitation to a temperature of 232 F. Deposit from funnel onto a bed of **Bitsyn®**. When cool, coat with milk chocolate.

Source: Pillsbury Co.

Chapter 5

CANDY COATINGS AND ICINGS

Coated candies retain their gloss, resist oxidative rancidity, do not bleed in hot weather, and, in the case of chocolate candies, resist blooming. Icings have much the same merits as coatings. Among the many varieties of coatings and icings are caramel coatings, chocolate coatings and icings, cocoa fondant coatings, milk hand icings, royal icing, and tropical icing.

Candy Coating for Apples

	% by Wt.
Sugar	59.30
Clinton 43 Bé regular corn syrup	39.50
Water	1.00
Salt	0.20
Color	to suit
Flavor	to suit
	Total 100.00

Procedure:

Cook sugar, corn syrup, and water to 285–290 F. Add color and salt and mix into batch thoroughly. Add imitation flavors, if desired.

Wash apples on sticks in advance and start dipping as soon as batch reaches proper temperature.

Source: ADM

Caramel Apple Coating

	% by Wt.
Water	2.72
Isomerose® 100 brand high fructose corn syrup	12.69
Sucrose	9.06
Clinton regular 43 Bé corn syrup	27.24
Sweetened condensed milk	45.33
Salt	0.21
Hardened vegetable oil*	2.72
Coloring	to suit
Bicarbonate of soda (F.C.C.)	0.03
	Total 100.00

Procedure:

Add ingredients in order shown above. Cook to a firm ball or about 250 F with stirring to prevent scorching.

Add bicarbonate of soda and stir thoroughly.

Have apples washed and on sticks in advance. Start dipping as soon as batch temperature drops to 220–230 F.

Source: ADM

Caramel Coating for Nougat Rolls

			% by Wt.
Corn syrup	30	lb	32.89
Sweetened condensed milk	25	lb	27.40
Hardened edible oil	4	lb	4.38
Nulomoline (invert sugar)	10	lb	10.96
Salt	4	oz	0.27
Granulated sugar	20	lb	21.91
Water	2	lb	2.19
Vanilla and butter flavors			to suit
			Total 100.00

* Coconut oil of approximately 96 F melting point is satisfactory. It may be necessary to vary the cook slightly according to weather conditions.

Procedure:

Place all of the ingredients, except the flavors, into a double-action mixing kettle, and mix well. Apply heat and cook the batch to a medium soft ball, then add the flavors.

This caramel may also be used for spreading to form the layer over nougat, coconut, etc., or nut meats may be mixed into the batch to form a caramel chew.

Source: Crompton & Knowles

Caramel for Enrobers

			% by Wt.
Corn syrup	10	lb	32.72
Nulomoline (invert sugar)	5	lb	16.36
Granulated sugar	5	lb	16.36
Hard coconut butter	8	oz	1.64
Sweetened condensed whole milk	10	lb	32.72
Salt	1	oz	0.20
		Total	100.00

Procedure:

Place corn syrup, **Nulomoline**, granulated sugar, and hard coconut butter together with water to dissolve sugar in a clean caramel cooking kettle and bring to a full rolling boil.

Add sweetened condensed milk to above solution and then cook to 227–228 F or a soft boil.

Add to cooked batch and mix well. Pour into enrober tank at once.

SUGGESTIONS:

Run only on enrober set up especially to handle caramel.

Source: Kraft General Foods Corp.

Caramel Popcorn

			% by Wt.
Sucrose (granulated cane or beet sugar)	7	lb	28.43
Brown sugar (medium)	4	lb	16.24
Water	3	lb	12.18
Clinton Nusweet "E" corn syrup (43 Bé)	10	lb	40.61
Salt	2	oz	0.51
Dairy butter, plastic vegetable butter, or coconut oil	8	oz	2.03
		Total	100.00

Procedure:

Weigh the proper amounts of sucrose, salt, and brown sugar into cooking kettle and add water. Start heating and stir until sugar is well dispersed. Add corn syrup and continue cooking until a temperature of 298–300 F is reached. Stir in butter, shortening, or oil just before adding to the popcorn. Stir sufficiently to prevent scorching. Pour directly onto popped corn in an oiled mixing kettle or in mixing machine and stir continuously until well blended. Pour onto oiled cooling slab until pieces are firm enough to pack.

If nonadhering pieces are desired, and to prevent formation of large corn and syrup chunks: (1) add syrup while very hot; (2) stir rapidly while mixing corn and syrup; (3) *spray or coat corn-brittle mixture with melted vegetable oil or coconut oil as soon as all syrup has been added and continue stirring mixture to keep chunks separate until the syrup has set*. Approximately $^{1}/_{4}$ lb of additional coconut oil or hydrogenated vegetable oil will be required for this purpose.

YIELD: Approximately 20 lb of finished popcorn syrup.

NOTE: The amount of popcorn to use will, of course, depend upon the amount of coating desired on the popcorn.

Source: ADM

Chocolate Summer Ice Coat

(For Hand or Machine Work)

			% by Wt.
PART 1—BASIC FONDANT (q.v.)	60	lb	42.81

PART 2:			
Water	25	lb	17.84
Cocoa powder (18% butter fat)	10	lb	7.13
Bicarbonate of soda	0.5	oz	0.02
PART 3:			
Granulated sugar	45	lb	32.12
Salt	1	oz	0.04
Vanilla flavor	1	oz	0.04
		Total	100.00

Procedure:

PART 1: Prepare basic fondant.

PART 2: Place the materials into a kettle, mix well, apply heat, and permit the batch to boil for a half minute.

PART 3: Add the granulated sugar to Part 2 and boil the batch to 226–230 F, then add the 60 lb of Basic Fondant (Part 1), plus the salt and vanilla flavor. Mix well until the fondant is melted, and the icing is ready for use.

Keep the icing warm and fluid when applying it.

NOTES:

This formula is used for coating cast cream centers, nougat, cast caramels, jellies, hard candies, fudge, coconut work. It may also be used for coating batches that should be later rolled in coconut, such as marshmallow, nougat, grained pulled candies, etc.

This coating when applied to heavy-type candies, such as cream centers, jellies, and caramels, will not spot. If marshmallows, grained nougat or other aerated candies are coated with this icing, spots will appear, and it is for this reason that after candies of this type are coated they should be rolled in coconut, nut meats, or sugar.

PART 1—BASIC FONDANT:		
Granulated sugar	80	lb
Corn syrup	20	lb
Nulomoline (invert sugar)	10	lb
Water (to dissolve the sugar)		

Procedure:

Heat altogether stirring the batch occasionally until it boils, then wash down all grains of sugar that adhere to the interior of the kettle and boil the batch rapidly to 244–246 F.

Source: Crompton & Knowles

Coating Machine Icing

			% by Wt.
Icing sugar (4X or 6X)	95	lb	52.18
Nulomoline (invert sugar)	5	lb	2.75
Gelatin (low grade)	2	lb	1.10
Water	40	lb	21.97
Hard coconut butter (90–92 F m.p.)	40	lb	21.97
Salt	1	oz	0.03
Flavor and color			to suit
			Total 100.00

Procedure:

Mix the gelatin with the water and heat to approximately 150 F. Place this into a marshmallow beater, add about half of the sugar and beat the batch well, then gradually add the remainder of the sugar and the **Nulomoline** and beat the batch until light.

Now add the coconut butter, which should be melted, then add color and flavor and beat well.

The icing should now be heated to approximately 110 F and run on the coating machine at temperatures ranging from 105 to 120 F.

If chocolate-flavored icing is desired, use 30 lb of the coconut butter and 12 lb of unsweetened chocolate melted together. The melted chocolate and hard coconut butter should be added after the batch has been beaten well.

When adding melted coconut butter or melted chocolate and coconut butter, the temperature of these products should range between 100 and 110 F.

Source: Crompton & Knowles

Cocoa Fondant Coating

			% by Wt.
PART 1—FLAVOR–COLOR BASE (q.v.)	10	lb	16.67
PART 2—FONDANT BASE (q.v.)	50	lb	83.33
			Total 100.00

COATING OPERATION:

Place 50 lb of the Fondant Base (Part 2) into a cream melting kettle, apply a little heat, mixing constantly, then add 10 lb of the Flavor–Color Base (Part 1) and continue to mix until the batch has been heated to approximately 135–145 F.

If the batch is too thick, add from 4–8 oz of water. Add vanilla or other flavors and mix well.

This coating may be used for bonbon centers of all types, including grained nougat, starch jelly gums, flexible marshmallow and other centers. This coating will also work satisfactorily on a standard cream fondant bonbon coating machine. The batch may be thinned down by the addition of water to secure a sufficiently fluid consistency for machine coating operation.

PART 1—FLAVOR–COLOR BASE:

Nulomoline (invert sugar)	10	lb
Plastic vegetable shortening (92 F m.p., as used by bakers)	10	lb
Hardened coconut oil (95–96 F m.p.)	5	lb
Salt	4	oz
Double dutched cocoa powder	10	lb
Natural process cocoa powder (18% butter fat)	5	lb

Procedure:

Heat the **Nulomoline** until melted (140–150 F). Add the plastic vegetable shortening and hardened coconut oil and mix until melted. Add the salt and sift in the cocoa powder, mixing until smooth. Place the batch into containers for future use.

PART 2—FONDANT BASE:

Granulated sugar	80	lb
Corn syrup	20	lb
Nulomoline (invert sugar)	10	lb
Water (to dissolve the sugar)		

Procedure:

Boil altogether to 242–244 F. Cool the melted syrup to approximately 120–130 F, then beat into fondant.

Source: Crompton & Knowles

Coconut Bonbon Coating

			% by Wt.
Fondant 90-10 (238–240 F cook)	100	lb	90.09
Nulomoline (invert sugar)	8	lb	7.21
Creamed Coconut	3	lb	2.70
Gelatin solution			to suit
			Total 100.00

Procedure:

Remelt fondant at 120–130 F. Add **Nulomoline, Creamed Coconut**, and gelatin solution. Mix well. Apply to centers at 120–130 F.

SUGGESTIONS:

Gelatin solution is made by dissolving 2 oz of gelatin in 4 oz of hot water.

Source: Kraft General Foods Corp.

Coconut Chocolate Pop Coating

			% by Wt.
Creamed Coconut	10	lb	9.98
Cocoa powder	10	lb	9.98
Sugar (6X)	20.75	lb	20.70
Skim milk powder	1.25	lb	1.25
Salt	8	oz	0.50
Coconut oil (76 F m.p.)	49.5	lb	49.36
Peanut oil	8	lb	7.98
Lecithin*	4	oz	0.25
			Total 100.00

* Add before grinding

Procedure:

Soften **Creamed Coconut** and add to dry ingredients. Mix well.

Run this paste over a 3 or 5 roll refiner. Grind to proper fineness. If paste is too dry adjust to proper consistency with coconut oil.

Blend ground paste thoroughly with liquid fats.

Pack in pail while hot. Store at temperature of 70 F or below.

Source: Kraft General Foods Corp.

Confectioners Hand Icing

(For Caramels, Nougat, Jellies, etc.)

			% by Wt.
Extra-X icing sugar	30	lb	73.96
Gelatin (100 bloom)	0.5	lb	1.23
Water (hot)	5	lb	12.33
Salt	1	oz	0.15
Frappé No. 1 (q.v.)	5	lb	12.33
Color and flavor			to suit
			Total 100.00

Procedure:

Mix the gelatin with all of the hot water, mixing well to completely dissolve the gelatin, then add the icing sugar and salt, mixing until smooth. Place the batch into a beater, add the frappé and beat until the batch is opaque. Add pink, yellow, or other color, with flavor to harmonize.

The hand coating is now ready to be applied to starch jelly gums, pectin, jellies, fudge, coconut work, chewing candies, hard candies, and plastic coconut centers. Three or four pounds of the centers are placed into a pan; then the operator dips both hands into the icing and rubs the icing over the centers, en masse. After the centers are thickly coated with icing, another operator takes a handful of coated centers and places them, one by one, on lightly oiled heavy waxed paper.

This icing will set rather quickly, and it will adhere to the center.

FRAPPÉ NO. 1:

Corn syrup	50	lb
Invert sugar	50	lb

Albumen	1 lb
Water	2 lb

Procedure:

Cook corn syrup to 245 F (sea level), shut off the heat and add invert sugar and stir until it is melted. Place the batch into a marshmallow beater, start the beater, then gradually add albumen which has been previously dissolved in water and beat the batch until light.

Source: Crompton & Knowles

Confectioners Icing

(For Coconut Squares, Caramels, Jellies, etc.)

		% by Wt.
Fine icing sugar	50 lb	68.49
Gelatin	1 lb	1.37
Hot water	12 lb	16.44
Frappé (q.v.)	10 lb	13.70
		Total 100.00

Procedure:

Mix the gelatin with the hot water and, if necessary, heat slightly until the gelatin is entirely dissolved. Place this into a marshmallow beating machine, preferably of the upright type. Add half of the sugar and beat for several minutes, then gradually add the balance of the sugar and beat until sufficiently light. Now add the frappé, color, and flavor.

Thin down with simple syrup to the required consistency.

It is customary for the operator to take a dipper full of this icing and mix it with 4 or 5 lb of the candies and then place them on wrapping paper to dry.

FRAPPÉ:

Corn syrup	50 lb
Nulomoline	50 lb
Albumen	1.5 lb
Water	3 lb

Procedure:

Cook the corn syrup to a temperature of 245 F. Shut off the steam and add the **Nulomoline** and stir until melted. Place the batch into a marshmallow beater, start the beater, then gradually add the albumen which has been dissolved in the water and beat until light.

Source: Crompton & Knowles

Confectioners Tropical Icing

(Hand Icing)

			% by Wt.
Molasses (**Grandma's**)	5	lb	6.94
Icing sugar	50	lb	69.45
Gelatin (low grade)	1	lb	1.39
Water	9	lb	12.50
Frappé No. 1 (q.v.)	7	lb	9.72
		Total	100.00

Procedure:

Heat the water and gelatin to approximately 150 F to dissolve the gelatin. Place the gelatin solution into a beater (upright beater preferred), then add half of the icing sugar and the molasses and beat for several minutes. Add the remainder of the icing sugar and frappé and beat until quite light.

Vanilla flavor may be added if desired.

FRAPPÉ NO. 1:

Corn syrup	50	lb
Nulomoline	50	lb
Albumen	1	lb
Water	2	lb

Procedure:

Boil the corn syrup to approximately 245 F. Shut off the heat, add the **Nulomoline**, and stir until melted. Place the batch into a beater, start the beater, then gradually add the albumen which has been dissolved in the water, and beat until light.

Source: Crompton & Knowles

Dark Compound Coating*

No. 1

	% by Wt.
Hardened vegetable fat (108 F) (hard butter)	24.19
Total fat	27.21
Cocoa matter (fat and moisture free)	15.12
Buttermilk powder	3.78
Sucrose	29.47
Lecithin	0.23
	Total 100.00

Viscosity @ 110 F — 55 ± 5° NCA
Flavor — to standard
Texture — smooth

Procedure:

Ingredients are mixed together in preliminary ingredient mix, followed by roll refining and conching. For conching, a time period not less than 6 h at 158–167 F is recommended. Tempering and enrobing should follow the recommendations of the supplier of the hard butter.

Source: RJR Nabisco

No. 2

	% by Wt.
Sugar	50.65
Palm kernel oil**	28.00
Natural cocoa powder (10–12%)	16.00
Dark dutch cocoa powder (10–12%)	4.00
Anhydrous butter oil (or **Spans** and **Tweens**)	1.00
Lecithin	0.25
Salt	0.05
Vanillin	0.05
	Total 100.00

* Chocolates and coatings should have a particle size not exceeding 25–35 μ to avoid a "sandy" mouthfeel.

** Could be replaced with regular hydrogenated P.K. or domestic nonlauric fat (Kaomel or Humkote).

Fineness — 0.006 in.
Total fat — 31–32%
Viscosity — 120±5 MacM @ 110 F

Procedure:
Conche 8–12 h @ 140 F

Source: Wilbur Chocolate Co.

Milk Compound Coating

	% by Wt.
Sugar	48.10
Palm kernel oil*	29.00
Natural cocoa (10–12%)	10.00
Whole spray-dried milk	6.50
Nonfat spray-dried milk	6.00
Lecithin	0.30
Salt	0.05
Vanillin	0.05
	Total 100.00

Fineness — 0.0006 in.
Total fat — 31–32%
Viscosity — 120±5 MacM @ 110 F

Procedure:
Conche: 8–12 h @ 140 F

Source: Wilbur Chocolate Co.

* Could be replaced with regular hydrogenated P.K. or domestic nonlauric fat (Kaomel or Humkote).

Dipping Caramel

			% by Wt.
Granulated sugar	12	lb	12.00
Corn syrup	20	lb	20.00
Nulomoline (invert sugar)	10	lb	10.00
Coconut oil (76 F)	2.5	lb	2.50
Sweetened condensed milk	42	lb	42.00
Creamed Coconut	3.25	lb	3.25
Fondant 80–20 (242 F)	10	lb	10.00
Salt	4	oz	0.25
			Total 100.00

Procedure:

Place granulated sugar, corn syrup, **Nulomoline**, and coconut oil together in a clean caramel kettle. Bring to a full rolling boil.

Now add milk slowly, $^1/_3$ at a time, allowing batch to come to a rolling boil after each addition. Finally, cook to a soft ball.

Turn off steam and add **Creamed Coconut**, fondant, and salt. Mix well and use for dipping while hot.

If caramel gets too thick, thin with warm simple syrup.

Source: Kraft General Foods Corp.

Dipping Marshmallow

			% by Wt.
Granulated sugar	20	lb	37.20
Corn syrup	20	lb	37.20
Water	6.25	lb	11.63
Gelatin	1.25	lb	2.34
Water	6.25	lb	11.63
			Total 100.00

Procedure:

Place sugar, corn syrup, and first amount of water together in kettle and cook to 230 F. Then pour in bowl of beater.

Dissolve gelatin in hot water (6.25 lb) and add to batch above. Beat up light and then use to dip chocolate bonbon centers.

If the above is too thick for dipping, thin with some hot sugar syrup made as follows:

Granulated sugar	16 lb
Water	8 lb 5 oz

Place together in kettle and bring to boil. Use for thinning while hot. Use no more than necessary to get proper fluidity in marshmallow.

SUGGESTIONS:

After bonbon centers have been dipped in this marshmallow, roll them in **Tender Fresh Macaroon** or **Tender Fresh Medium Coconut**. This makes the finished Snow Ball.

Source: Kraft General Foods Corp.

Fondant for Dipping

			% by Wt.
Granulated sugar	85	lb	65.66
Corn syrup	15	lb	11.59
Nulomoline (invert sugar)	7	lb	5.41
Water	20	lb	15.45
Hard coconut butter (80–90 F m.p.)	2	lb	1.55
Medium grade gelatin (soaked in the water)	2	oz	0.10
Cold water	5	oz	0.24
			Total 100.00

Procedure:

Place water, sugar, **Nulomoline**, and corn syrup into a kettle and bring the batch to boiling point. Stir occasionally so as to be certain that sugar is dissolved when the batch begins to boil. Wash down sides of the kettle to remove all grains of sugar. Place a thermometer into batch and cook to 244–248 F.

Turn off heat, add the gelatin solution and mix the batch gently until gelatin is dissolved. Cool batch to approximately 105 F, add coconut butter which should be melted, then beat into fondant. It is advisable to use the fondant the same day on which it is made.

If this fondant is to be used in a coating machine, it is suggested that it be boiled to 238 F, then cooled to 100 F and beaten in the usual manner.

When preparing the fondant for machine work, an additional 5 lb of **Nulomoline** should be added to 100 lb of fondant which should be heated to 105–110 F.

Source: Crompton & Knowles

Icing for Hand or Machine Work

		% by Wt.
Sugar (XXXX)	50 lb	68.49
Nulomoline (invert sugar)	3 lb	4.11
Water	10 lb	13.70
Frappé No. 1 (q.v.)	10 lb	13.70
		Total 100.00

Procedure:

For icing to be run on the coating machine, place 7 lb of water into a kettle, add half of the sugar, the **Nulomoline**, and frappé and mix thoroughly, then add the balance of the sugar and beat for 2 or 3 min or until the batch is smooth.

Heat the icing to approximately 120 F for machine work.

For hand icing, place 10 lb of water into a kettle and continue as for machine work. When applying the icing by hand, do not heat the batch; simply beat for several minutes, adding color, flavor, etc.

This icing may be used on marshmallows, caramels, nougat, coconut, and jellies.

FRAPPÉ NO. 1:

Corn syrup	50 lb
Nulomoline (invert sugar)	50 lb
Albumen (dissolved in the water)	1 lb
Water	2 lb

Procedure:

Boil the corn syrup to approximately 245 F, shut off the heat and add the **Nulomoline** and stir until it is melted. Place the batch into a beater, start the beater, then gradually add the albumen solution and beat the batch until light.

Source: Crompton & Knowles

Jelly for Coating

(Marshmallow, Coconut Work, Cast Cream Centers, Nougat, etc.)

			% by Wt.
Granulated sugar	45	lb	44.55
Nulomoline (invert sugar)	15	lb	14.85
Agar agar (Japanese gelatin)	1	lb	1.00
Water	40	lb	39.60
			Total 100.00

Procedure:

Soak the agar agar in the 40 lb of water for at least 1 h, then gradually bring it to the boiling point and strain through a #30 sieve.

Add the sugar and **Nulomoline** and boil the batch to 222 F (sea level).

The coating may be applied either hot or at a temperature of around 125 F, in order to assure maximum coverage of the candies.

Flavors or colors containing acid should not be added to the jelly until it has cooled to 130 F or less.

This coating may be applied by hand or machine when coating cast cream centers, grained nougat, cast or cut marshmallow, coconut balls, etc.

Source: Crompton & Knowles

Milk Hand Icing

			% by Wt.
Sugar (XXXX)	45	lb	55.19
Nulomoline (invert sugar)	3	lb	3.67
Sweetened condensed whole milk	15	lb	18.39
Frappé No. 1	9	lb	11.03
Gelatin (100 bloom)	9	oz	0.69
Water (hot)	9	lb	11.03
			Total 100.00

Procedure:

Soak the gelatin in the hot water, then add the **Nulomoline**, half of the sugar and beat light. Then add the balance of the sugar and the condensed milk, mix well, then add the frappé, color, flavor, and mix well.

FRAPPÉ NO. 1:

Corn syrup	50	lb
Invert sugar	50	lb
Albumen	1	lb
Water	2	lb

Procedure:

Cook corn syrup to 245 F (sea level), shut off the heat and add invert sugar and stir until it is melted. Place the batch into a marshmallow beater, start the beater, then gradually add albumen which has been previously dissolved in water and beat the batch until light.

Source: Crompton & Knowles

Royal Icing

			% by Wt.
Powdered sugar	79	lb	78.99
Nulomoline (invert sugar)	10	lb	10.00
Egg whites	10	lb	10.00
Water	1	lb	0.99
Acetic acid sol'n. (33.3%)	0.25	oz	0.02
		Total	100.00

Procedure:

Add the water to the **Nulomoline** and heat until it is a clear syrup. The temperature should be about 150 F. Put this into a beater with the egg whites and add about half of the sugar and beat until fairly light, then add the balance of the sugar and beat until stiff enough so that it will hold its shape. While beating add the acetic acid solution.

Source: Crompton & Knowles

Special Bonbon Coating

			% by Wt.
Special Fondant (q.v.)	10	lb	83.33
Edible oil (78 F m.p.)	1	lb	8.33

Frappé No. 2 (q.v.)	1	lb	8.34
Flavor and color			to suit
			Total 100.00

Procedure:

Place the fondant into a double boiler, add the edible oil and continue to mix until the fondant is melted and heat to approximately 120–130 F. Now add the frappé, flavor, and coloring, and mix thoroughly.

If necessary, add water to reduce the batch to the desired consistency. The fondant may be heated to higher temperatures if quicker setting action is desired.

SPECIAL FONDANT:

Granulated sugar	40	lb
Nulomoline (invert sugar)	5	lb
Corn syrup	5	lb
Water (to dissolve sugar)		

Procedure:

Cook the batch to 244 F. Cool the resulting syrup to approximately 100 F, then beat into fondant.

FRAPPÉ NO. 2:

Corn syrup	25	lb
Nulomoline (invert sugar)	25	lb
Albumen (dissolved in the water)	1	lb
Cold water	2	lb

Procedure:

Boil the corn syrup to 245 F. Turn off the heat, add the **Nulomoline** and stir until it is melted. Place the batch into a marshmallow beater, start the beater, then gradually add the albumen solution and beat the batch until light.

Source: Crompton & Knowles

Vanilla Icing

(For Machine Work)

			% by Wt.
Refined granulated sugar	100	lb	48.54
Nulomoline (invert sugar)	25	lb	12.13
Water	25	lb	12.13
Gelatin (200 bloom)*	1	lb	0.50
Hard coconut butter (90 F m.p.)	30	lb	14.56
Pulverized sugar	25	lb	12.14
			Total 100.00

Procedure:

Place the granulated sugar, **Nulomoline**, and water into a clean kettle and bring the batch to the boiling point. Wash down the sides of the kettle to remove all grains of sugar. Place a tested thermometer into the batch and boil to 230 F. Add the soaked gelatin and stir until it is dissolved. Permit the batch to cool to approximately 160 F. Place the batch into a beating machine and beat until it is quite white and grain is visible, then add the coconut butter (melted). Beat well, then work in the pulverized sugar and beat thoroughly.

Add water to secure the proper consistency for running on the machine.

Maintain a temperature of approximately 115 F when running this icing.

Source: Crompton & Knowles

Winter Coconut Coating

				% by Wt.
Creamed Coconut	20	lb 13.5	oz	20.81
Hard butter	26	lb 0.5	oz	26.26
Sugar (6X or microfine)	44	lb 5.5	oz	44.26
Whole milk powder	8	lb 11.0	oz	8.67
Salt				to suit
				Total 100.00

* Soaked in 10 lb of cold water for at least 15 min.

Procedure:

Melt **Creamed Coconut** and hard butter, mix and heat to 120 F.

Mix together and add to the above; heat if necessary to keep mass fluid. Blend well in vertical mixer.

Temper with stirring to 88 F, pour into mold then cool to 35 F.

SUGGESTIONS:

This coating can be made on regular chocolate coating machinery, handling the **Creamed Coconut** as though it were chocolate and the hard butter as though it were cocoa butter.

Source: Kraft General Foods Corp.

Chapter 6

COCONUT CANDIES

Candy Coconut Cakes

(Brown on Top Only)

			% by Wt.
Corn syrup	50	lb	33.28
Nulomoline (invert sugar)	10	lb	6.66
Granulated sugar	40	lb	26.62
Desiccated macaroon coconut	30	lb	19.97
White corn meal	20	lb	13.30
Salt	4	oz	0.17
Vanilla flavor			to suit
			Total 100.00

Procedure:

Place into a steam kettle the corn syrup, **Nulomoline,** 10 lb of water, and the sugar; bring the batch to boiling point, then add the coconut and cook batch to a medium firm paste. Now add the corn meal, salt, and flavor, and mix thoroughly. This batch should be cooked firm so that it will be possible to roll it into sheets and then cut into cakes of the desired size.

After the batch has been cooked, spread it onto a cooling slab which has been oiled and lightly dusted with flour. Roll the batch down to desired thickness and after the batch is cold, paint the top of the batch with the following solution:

Granulated sugar	10	lb
Water	3	lb
Molasses (**Grandma's**)	1	pt

Albumen	1/2 oz
Water	3 oz

Procedure:

Bring the sugar and water to the boiling point. Add the molasses. Allow this syrup to cool to at least 100 F, then add 1/2 oz of albumen which has been dissolved in 3 oz water.

After the sheets of coconut have been painted with the above solution they may then be placed into an oven of the grill type which is equipped with a top flame only as it is essential that the bottom is not baked. After sheets have been browned on the surface, they may be cut into the desired size.

Source: Crompton & Knowles

Cast Coconut Caramel

		% by Wt.
Sugar	31 lb	24.31
Corn syrup	40 lb	31.36
Invert sugar	10 lb	7.84
Creamed Coconut	30 lb	23.52
Lecithin	0.5 oz	0.02
Water	4 lb	3.14
Sweetened condensed milk	12 lb	9.41
Salt	4 oz	0.20
Vanilla extract	4 oz	0.20
		Total 100.00

Procedure:

Place the sugar and corn syrup, invert sugar, **Creamed Coconut**, lecithin, water, and sweetened condensed milk in a caramel cooking kettle. Cook to a medium ball.

Add salt and vanilla. Mix well. Cast into starch. When set, shake out and dip in chocolate.

Source: Kraft General Foods Corp.

Cherry Coconut Creams

			% by Wt.
Gem Macaroon Coconut	18	lb	11.35
Invert sugar	4	lb	2.52
Water (hot)	2	lb	1.26
Salt	4	oz	0.16
Maraschino cherries	2	lb	1.26
Sugar	100	lb	63.07
Invert sugar	12	lb	7.57
Invertase	2	oz	0.08
Creamed Coconut	5	lb	3.15
Vanilla flavor	3	oz	0.12
Nougat cream	15	lb	9.46
Pink color, maraschino flavor			to suit
			Total 100.00

Procedure:

Mix water with salt and 4 lb invert sugar. Add **Gem Macaroon Coconut**. Mix thoroughly. Then spread on open-type fondant beater with chopped, drained maraschino cherries.

Place sugar and invert sugar in kettle with enough water to dissolve the sugar. Bring to boil, work down sugar crystals from sides of kettle. Cook to 244–246 F. Pour cooked batch over prepared coconut on fondant beater. Sprinkle surface with water. Cool to 100 F, add **Creamed Coconut** cut in small pieces. Start beater, add invertase and vanilla flavor, pink color, and maraschino flavor. When batch shows grains add nougat cream.

SUGGESTIONS:

After beating allow batch to set for 5–10 min. Then knead until soft after which it is ready to form on a Friend machine or by hand.

Source: Kraft General Foods Corp.

Coconettes

			% by Wt.
PART 1:			
Desiccated coconut	30	lb	28.84
Water	6	lb	5.77
Nulomoline (invert sugar)	8	lb	7.69

PART 2:		
Corn syrup	15 lb	14.42
Granulated sugar	25 lb	24.04
Water (to dissolve sugar)		
PART 3:		
Basic Fondant (q.v.)	10 lb	9.62
Ground preserved pineapple, fruit peels, nuts, etc.	5 lb	4.81
Frappé No. 1 (q.v.)	5 lb	4.81
Color, flavor		to suit
		Total 100.00

Procedure:

PART 1: Mix the water with the **Nulomoline**, add the coconut and mix well; then set aside for 10 or 15 min.

PART 2: Boil altogether to 255–260 F (sea level).

Pour this on top of the freshened coconut (Part 1), mix well, then add Part 3.

Mix the batch thoroughly, then spread out on waxed paper and when cold cut into bars or pass the batch through a rolled center machine. Coat with chocolate or icing, or place into paper cups.

BASIC FONDANT:	
Granulated sugar	80 lb
Corn syrup	20 lb
Nulomoline	10 lb
Water (to dissolve the sugar)	

Procedure:

Boil the granulated sugar, corn syrup, **Nulomoline** together with sufficient water to dissolve the sugar to 240–244 F, then cool and beat into fondant in the usual manner.

FRAPPÉ NO. 1:	
Corn syrup	50 lb
Invert sugar	50 lb
Albumen	1 lb
Water	2 lb

Procedure:

Cook corn syrup to 245 F (sea level), shut off the heat and add invert sugar and stir until it is melted. Place the batch into a marshmallow beater, start the beater, then gradually add albumen which has been previously dissolved in water and beat the batch until light.

Source: Crompton & Knowles

Coconut Bonbon Center

No. 1

			% by Wt.
Gem Medium Coconut	50	lb	33.22
Nulomoline (invert sugar)	10	lb	6.64
Corn syrup	50	lb	33.22
Sugar	25	lb	16.61
Creamed Coconut	15	lb	9.97
Salt	4	oz	0.17
Vanilla	4	oz	0.17
			Total 100.00

Procedure:

Place **Gem Medium Coconut** and **Nulomoline** in a stirring machine and let mix at low speed for 15 min.

Cook corn syrup and sugar with water to dissolve to 228 F.

Now add **Creamed Coconut** and salt to cooked syrup. This makes a coconut-flavored syrup. Pour this syrup over the freshened **Gem Medium Coconut** in the stirring machine. Add vanilla and mix well.

Source: Kraft General Foods Corp.

No. 2

			% by Wt.
Gem Medium Coconut	12	lb	33.68
Nulomoline (invert sugar)	1	lb	2.80
Water	8	oz	1.40

Corn syrup	12	lb	33.68
Sugar	6	lb	16.85
Creamed Coconut	3	lb	8.42
Frappé (q.v.)	1	lb	2.81
Salt	1	oz	0.18
Vanilla extract	1	oz	0.18
			Total 100.00

Procedure:

Place **Gem Medium Coconut, Nulomoline,** and water in mixer and let mix at low speed for 10–15 min.

Cook corn syrup and sugar with water to dissolve to 230 F.

Add **Creamed Coconut**, frappé, salt, and vanilla extract.

NOTE:

This center may be dipped in fondant or in marshmallow.

FRAPPÉ:

Corn syrup	5	lb
Nulomoline	5	lb
Cold water	1.5	oz

Procedure:

Place the corn syrup in kettle and cook to 245 F. Turn off heat and add the **Nulomoline,** stir until melted. Place batch in a marshmallow beater and start to beat. Then gradually add the cold water.

Source: Kraft General Foods Corp.

No. 3

					% by Wt.
Fondant 90-10 (q.v.)	80	lb	10.75	oz	80.85
Syrup 90-10 (q.v.)	16	lb	1.75	oz	16.14
Creamed Coconut	3	lb			3.01
Flavor, color					to suit
					Total 100.00

Procedure:

Remelt fondant in kettle at 106 F and mix well with syrup and **Creamed Coconut.**

Add flavor and color and place in enrober and run at temperature of approximately 104 F.

FONDANT 90-10:

Sugar	90 lb
Corn syrup	10 lb
Water (to dissolve sugar)	

Procedure:

Place sugar with corn syrup in kettle with water to dissolve sugar. Cook to 242 F. Cool to 110 F and beat into fondant.

SYRUP 90-10:

Sugar	90 lb
Corn syrup	10 lb
Water (to dissolve sugar)	

Procedure:

Place the sugar with corn syrup in kettle with water to dissolve sugar. Warm to melt sugar.

Source: Kraft General Foods Corp.

No. 4

			% by Wt.
Nulomoline (invert sugar)	30	lb	16.18
Water	10	lb	5.39
Desiccated macaroon coconut	75	lb	40.45
Corn syrup	70	lb	37.74
Salt	7	oz	0.24
			Total 100.00

Procedure:

Heat the **Nulomoline** and water to approximately 150 F. Add the coconut, mix thoroughly and permit the batch to stand for at least 1 h.

Heat the corn syrup until quite hot (do not boil); then add the corn syrup and salt to the batch, mix thoroughly and spread out to cool.

Source: Crompton & Knowles

Coconut Bonbon Fondant

			% by Wt.
Fondant 80-20 (cooked to 242–244 F)	100	lb	90.29
Creamed Coconut	5	lb	4.51
Simple syrup (2:1 sugar:water)	5.5	lb	4.97
Gelatin sol'n. (5 oz gelatin dissolved in10 lb water)	4	oz	0.23
		Total	100.00

Procedure:

Place fondant in remelt kettle and melt at 145 F.

Add **Creamed Coconut**, simple syrup, and gelatin solution. Mix well. Cook to 122 F and dip.

Source: Kraft General Foods Corp.

Coconut Butter Cast Creams

			% by Wt.
Sugar	100	lb	47.45
Corn syrup	55	lb	26.10
Fondant 70-30 (cooked to 244 F)	15	lb	7.12
Frappé	30	lb	14.23
Creamed Coconut	10	lb	4.74
Salt	4	oz	0.12
Butter flavor	6	oz	0.18
Vanilla flavor	2	oz	0.06
		Total	100.00

Procedure:

Add water to sugar and corn syrup to dissolve. Cook to 265 F. When batch stops boiling, add fondant and stir until batch shows signs of grain.

Then add frappé, **Creamed Coconut**, salt, and flavors. Mix well and cast into dry starch. Shake out and brush very clear. Glaze in revolving pans as desired.

NOTE:

Frappé should have $1^1/_2$ lb of egg albumen per 100 lb of frappé.

Source: Kraft General Foods Corp.

Coconut Butterscotch Kisses

			% by Wt.
Corn syrup	26	lb	38.42
Granulated sugar	22	lb	32.50
Creamed Coconut	6	lb	8.86
Salted dairy butter	2.5	lb	3.69
Invert sugar	4	lb	5.91
Cream (20%)	4	lb	5.91
Frappé (very light)	3	lb	4.43
Flavor	3	oz	0.28
			Total 100.00

Procedure:

Place corn syrup, granulated sugar, **Creamed Coconut**, salted dairy butter, invert sugar, and 20% cream together in a kettle with enough water to dissolve sugar and cook to a firm ball.

When batch stops boiling, add frappé, salt, and flavor. Cool to handle. Do not pull. When batch has cooled sufficiently, spin, cut, and wrap.

Source: Kraft General Foods Corp.

Coconut Cakes and Kisses

(Plastic)

			% by Wt.
PART 1:			
Desiccated coconut (part medium shred, part macaroon)	15	lb	31.23

Water	3 lb	6.25
Nulomoline (invert sugar)	4 lb	8.32
PART 2:		
Sugar	15 lb	31.23
Corn syrup	5 lb	10.41
PART 3:		
Short fondant	5 lb	10.41
Sucrovert	0.5 oz	0.07
Frappé No. 1 (q.v.)	1 lb	2.08
Color and flavor		to suit
		Total 100.00

Procedure:

PART 1: Mix the water with the **Nulomoline**, add the coconut, mix well and set aside for about 15 min.

PART 2: Boil altogether to 255–260 F.

Pour this into the freshened coconut (Part 1); mix well, add short fondant and **Sucrovert** and mix thoroughly. Then add Frappé No. 1, color and flavor as desired, and mix well.

Spoon out into kisses, cakes, or bars.

FRAPPÉ NO. 1:

Corn syrup	50 lb
Invert sugar	50 lb
Albumen	1 lb
Water	2 lb

Procedure:

Cook corn syrup to 245 F (sea level), shut off the heat and add invert sugar and stir until it is melted. Place the batch into a marshmallow beater, start the beater, then gradually add albumen which has been previously dissolved in water and beat the batch until light.

Source: Crompton & Knowles

Coconut Caramel

			% by Wt.
Sweetened condensed milk	50	lb	39.45
Corn syrup	50	lb	39.45
Salt	4	oz	0.19
Creamed Coconut	15	lb	11.84
Fondant 80-20	11.5	lb	9.07
Color			to suit
			Total 100.00

Procedure:

Place sweetened condensed milk, corn syrup, and salt in a kettle and cook to medium ball.

Add **Creamed Coconut**, fondant, and color. Mix well. Spread on greased slab to about 1/4 in. When cool, add coconut center.

FONDANT 80-20:

Sugar	80	lb
Corn syrup	20	lb
Water (to dissolve sugar)		

Procedure:

Cook to 244 F.

Source: Kraft General Foods Corp.

Coconut Caramel Swirls

			% by Wt.
CARAMEL BATCH:			
Corn syrup	20	lb	29.25
Sugar	15	lb	21.95
Sweetened condensed milk	17	lb	24.86
Hard butter (88–90 F)	13	lb	19.01
Creamed Coconut	3 lb 5	oz	4.84
Salt	1	oz	0.09
			Total 100.00

CENTER BATCH:

Corn sugar	16 lb	53.68
Sugar	6.5 lb	21.80
Nougat cream	3 lb	10.06
Creamed Coconut	1 lb	3.35
Gem Macaroon Coconut	3 lb 5 oz	11.11
		Total 100.00

Procedure:

CARAMEL BATCH: Place corn syrup, sugar, sweetened condensed milk, and hard butter in kettle with water to dissolve and cook to soft caramel. Add **Creamed Coconut**, salt, and flavor. Mix well, then spread on oiled slab to $^{1}/_{4}$ in. or less thickness.

CENTER BATCH: Place corn syrup, and sugar in kettle with water to dissolve. Cook to firm ball (250 F). Add nougat cream, **Creamed Coconut** and **Gem Macaroon Coconut**. Mix well. Spread on caramel batch and allow to cool. Roll up in jelly-roll shape and pull down to 1 in. thickness. Cut and wrap on kiss machine.

SUGGESTIONS:

Caramel batch can be made in three flavors: vanilla, chocolate, or maple. To make chocolate caramel substitute half of **Creamed Coconut** with chocolate liquor or cocoa powder and add artificial chocolate color if darker color is desired.

Source: Kraft General Foods Corp.

Coconut Chocolate Fudge

			% by Wt.
Granulated sugar	2 lb	8 oz	18.10
Invert sugar		8 oz	3.62
Sweetened condensed milk	1 lb	4 oz	9.05
Evaporated milk	1 lb		7.24
Chocolate liquor	1 lb		7.24
Gem Medium Coconut	1 lb		7.24
Fondant	6 lb	9 oz	47.51
Salt, vanilla			to suit
			Total 100.00

Procedure:

Place all ingredients except salt and vanilla in an agitator equipped kettle. Cook to 244–245 F. Cool to 200 F.

Add fondant, salt, and vanilla. Mix well. Pour out, allow to set, cut, and enrobe.

FONDANT:		
Granulated sugar	5	lb
Invert sugar	1	lb
Water (to dissolve sugar)		
Creamed Coconut	9	oz

Procedure:

Combine sugar, invert sugar, and water to dissolve. Cook to 245 F. Place in cream beater. Cool to 100 F. Start to cream when batch shows grain. Add **Creamed Coconut**. Finish creaming.

Source: Kraft General Foods Corp.

Coconut Cream

					% by Wt.
Medium desiccated coconut	8	lb			11.38
Hot water*			11	oz	0.98
Invert sugar*	1	lb	8	oz	2.13
Amerfond fondant sugar	50	lb			71.11
Powdered, spray-dried egg albumen			4	oz	0.36
Regular corn syrup	5	lb	4	oz	7.47
Hot water	4	lb	8	oz	6.40
Invertase			2	oz	0.17
Coconut flavoring					to suit
					Total 100.00

Procedure:

Mix together invert sugar and hot water; then stir this into the coconut and set to one side until required.

* For moistening coconut

Place half of the fondant sugar in the bowl of a planetary mixer, and, at low speed, dry mix in the powdered albumen.

Add corn syrup and water and mix thoroughly; then turn the mixer to high speed and beat for 4 min.

Turn mixer to low speed, add the balance of the fondant sugar, invertase, and coconut flavoring and mix well; then mix in the moistened coconut.

Allow batch to stand until firm.

Roll or extrude and coat with chocolate or pastel coating.

Source: Amstar

Coconut Cream Bars and Eggs

(Cast into Starch)

			% by Wt.
PART 1—FONDANT (q.v.)	75	lb	48.35
PART 2—BOB:			
Sugar	45	lb	29.00
Corn syrup	5	lb	3.22
Water	10	lb	6.45
PART 3:			
Standard Frappé (q.v.)	10	lb	6.45
Convertit	2	oz	0.08
Vanilla flavor			to suit
PART 4—FRESHENED COCONUT (q.v.)	10	lb	6.45
			Total 100.00

Procedure:

PART 1: Prepare fondant.

PART 2: Boil bob altogether to 236–240 F.

Pour bob into the fondant (Part 1), mix well, then add the Standard Frappé and mix well. Now add **Convertit** and while mixing add vanilla flavor and the freshened coconut (Part 4). Mix well and cast into starch. Casting temperature 150–155 F.

FONDANT:

Sugar	80 lb
Corn syrup	10 lb
Nulomoline (invert sugar)	10 lb
Water	22 lb

Procedure:

Boil altogether to 242–244 F, then cool and beat into fondant in the usual manner.

STANDARD FRAPPÉ:

Corn syrup	50 lb
Nulomoline (cold)	50 lb
Albumen	1 lb
Water	2 lb

Procedure:

Boil the corn syrup to 245 F. Shut off steam and add the cold **Nulomoline** and stir until it is melted. Place batch into a marshmallow beater, start the beater then add albumen which has been previously dissolved in water; beat until light.

FRESHENED COCONUT:

Water	2.5 lb
Nulomoline (invert sugar)	5 lb
Fine desiccated coconut	20 lb

Procedure:

Heat the **Nulomoline** and water to approximately 150 F, add the coconut, mix well, and allow to set for half an hour before adding to the batch.

Source: Crompton & Knowles

Coconut Cream Cheese Fudge

		% by Wt.
Cream cheese	25 lb	25.00
Sugar (6X)	5 lb	5.00

Chocolate coating (dark or milk)	50	lb	50.00
Tender Fresh Medium Coconut	20	lb	20.00
			Total 100.00

Procedure:

Cream together cream cheese and 6X sugar.

Melt chocolate coating and add to cream cheese–sugar mixture.

Add **Tender Fresh Medium Coconut** and mix in quickly, but gently. Do not overmix.

Pour out on slightly greased slabs. When set, cut.

Source: Kraft General Foods Corp.

Coconut Cream Patties

			% by Wt.
Gem Medium Coconut	24	lb	10.53
Water	3	lb	1.32
Invert sugar	6	lb	2.63
Sugar	60	lb	26.32
Water	20	lb	8.77
Corn syrup	10	lb	4.39
Fondant (q.v.)	100	lb	43.85
Frappé (q.v.)	5	lb	2.19
Vanilla			to suit
			Total 100.00

Procedure:

Place **Gem Medium Coconut** in mixer. Add 3 lb of water and melted invert sugar. Blend.

Combine sugar, 20 lb of water, and corn syrup and cook to 242 F.

Add fondant and frappé to bob and mix until dissolved. Lastly, add freshened coconut and mix well.

Cast into dry starch molds. Add **Golden Toasted Flake Coconut** or peanuts. Enrobe in chocolate.

FRAPPÉ:

Corn syrup	80	lb
Sugar	20	lb
Water	1	gal
Tapioca flavor	4	lb
or		
Cooking starch (suspended in water)	5	lb
Water	2	gal
Albumen	3	lb
Cold water	2	qt

Procedure:

Bring the corn syrup, sugar, and 1 gal water to boil, then add tapioca flavor or cooking starch suspended in 2 gal of water. Cook to 230 F. Pour into beater. Beat until temperature is 160 F or lower. Add albumen dissolved in 2 qt cold water. Beat until light.

FONDANT:

Sugar	80	lb
Water	28	lb
Corn syrup	40	lb
Creamed Coconut	6	lb

Procedure:

Cook sugar and water to 235 F. Add corn syrup and **Creamed Coconut**. Cook to 120 F. Cream.

Source: Kraft General Foods Corp.

Coconut Cream Wafers

			% by Wt.
Gem Macaroon Coconut	3	lb	3.03
Nulomoline (invert sugar)	8	oz	0.50
Water	8	oz	0.50
Sugar	72	lb	72.72
Water	16 lb 15	oz	17.10

Nulomoline (invert sugar)	3	lb	3.03
Creamed Coconut	3	lb	3.03
Gelatin (dissolved)	1/2	oz	0.03
Hot water	1	oz	0.06
			Total 100.00

Procedure:

Place **Gem Macaroon Coconut**, first portion of **Nulomoline**, and 8 lb of water in mixer. Mix well for 5–10 min.

Place sugar, water, **Creamed Coconut**, and second portion of **Nulomoline** in kettle and cook to 242–245 F.

Turn off steam and add gelatin. Stir gelatin gently into batch. Pour in moistened open-type fondant beater. Sprinkle surface of batch with water. Cool to 100 F and beat until firm.

Place fondant in remelt kettle and heat until melted. Add color and flavor as desired. Lastly add tenderized coconut. Stir well. Check temperature to 140–150 F.

Coconut fondant may be cast on rubber sheets, molds, or waxed paper. Dry bottoms before packing.

Source: Kraft General Foods Corp.

Coconut Crisp

			% by Wt.
Corn syrup	6	lb	38.25
Sugar	6	lb	38.25
Salt	2	oz	0.80
Coconut oil	1	oz	0.40
Gem Slice Coconut (hot)	2.5	lb	15.93
Gem Coarse Coconut (hot)	1	lb	6.37
			Total 100.00

Procedure:

Cook corn syrup, sugar, salt, and coconut oil to 290 F.

Add hot **Gem Slice** and **Gem Coarse Coconut** to syrup and mix well. Dump slab and spread out. Cut into bars or allow to cool and break up as brittle.

SUGGESTIONS:

If lighter crisp is desired, add 1 oz bicarbonate of soda to hot syrup just before adding coconut. Add caramel color if syrup is not dark enough in color.

Source: Kraft General Foods Corp.

Coconut Glaze

					% by Wt.
Granulated sugar	20	lb	5	oz	20.30
Corn syrup	10	lb	2	oz	10.12
Water	8	lb	2	oz	8.12
Condensed milk	4	lb	1	oz	4.06
Creamed Coconut	6	lb	3	oz	6.18
Lecithin			1	oz	0.06
Salt			3	oz	0.19
Vanilla extract			3	oz	0.19
Gem Slice Coconut	40	lb	10	oz	40.60
Black walnuts	10	lb	3	oz	10.18
					Total 100.00

Procedure:

Place sugar, syrup, and water in kettle and bring to full, rolling boil.

Add condensed milk and cook batch to 290 F.

Remove from fire and add **Creamed Coconut**, lecithin, salt, and vanilla extract. Mix well.

Now add **Gem Slice Coconut** and walnuts which have been preheated in an oven. Mix well. Spread out thin on a greased slab and when cool, break up.

Source: Kraft General Foods Corp.

Coconut Hard Candy Center

			% by Wt.
Gem Medium Coconut	2	lb	8.74
Invert sugar	8	oz	2.19
Water	4	oz	1.09
Fondant 80-20	10	lb	43.72

Creamed Coconut	1	lb	4.37
Sugar	6	lb	26.23
Corn syrup	1	lb	4.37
Frappé (q.v.)	2	lb	8.74
Convertit (invertase)	2	oz	0.55
Color and flavor			to suit
			Total 100.00

Procedure:

Place **Gem Medium Coconut**, melted invert sugar, and water in mixer. Blend well.

Place fondant, **Creamed Coconut**, and tenderized **Gem Medium Coconut** in remelt kettle.

Combine sugar, corn syrup, and water to dissolve sugar in kettle. Cook to 244 F. Pour syrup over fondant and **Creamed Coconut**.

Lastly add frappé, **Convertit**, color, and flavor. Mix well.

Center may be used for caramels or hard candy.

FRAPPÉ:

Sugar	20	lb
Corn syrup	40	lb
Egg albumen	1	lb
Cold water	2	lb

Procedure:

Cook sugar to 232 F. Add the corn syrup and place in beater. Add the egg albumen which has been dissolved in water. Beat to a dry beat.

Source: Kraft General Foods Corp.

Coconut Haystacks

No. 1

			% by Wt.
Gem Strip Coconut	8	lb	9.16
Water	2	lb	2.29
Invert sugar	2	lb	2.29

Granulated sugar	10	lb	11.44
Brown sugar	5	lb	5.72
Corn syrup	20	lb	22.89
Sweetened condensed milk	12	lb	13.74
Creamed Coconut	4	lb	4.58
Invert sugar	4	lb	4.58
Salt	3	oz	0.21
Fondant 80-20 (q.v.)	20	lb	22.89
Vanilla flavor	3	oz	0.21
Color			to suit
			Total 100.00

Procedure:

Place **Gem Strip Coconut** and water in mixer. Blend well. Add melted invert sugar. Mix well.

Combine granulated sugar, brown sugar, corn syrup, sweetened condensed milk, and **Creamed Coconut** together in an agitator-equipped steam kettle. Cook to 248–250 F.

Add invert sugar, fondant, and vanilla. Stir well. Lastly add tenderized **Gem Strip Coconut** and mix well. Deposit as haystacks.

FONDANT 80-20:

Sugar	80	lb
Corn syrup	20	lb

Procedure:

Cook to 244 F.

Source: Kraft General Foods Corp.

No. 2

			% by Wt.
Gem Medium Coconut	28	lb	26.07
Water	2	lb	1.86
Invert sugar	2	lb	1.86
Granulated sugar	10	lb	9.31
Brown sugar	5	lb	4.66
Corn syrup	20	lb	18.63

Sweetened condensed milk	12	lb	11.18
Creamed Coconut	4	lb	3.73
Salt	3	oz	0.17
Invert sugar	4	lb	3.73
Fondant 80-20	20	lb	18.63
Vanilla flavor	3	oz	0.17
Color			to suit
			Total 100.00

Procedure:

Place **Gem Medium Coconut** and water in mixer and blend well. Add melted invert sugar and mix well.

Place granulated sugar, brown sugar, corn syrup, sweetened condensed milk, and **Creamed Coconut** together in an agitator-equipped steam kettle. Cook to 248–250 F.

Add salt, invert sugar, fondant, and vanilla to cooked syrup. Stir well. Lastly, add tenderized **Gem Medium Coconut** and mix well. Deposit as haystacks.

FONDANT:

Sugar	80	lb
Corn syrup	20	lb

Procedure:

Cook to 244 F.

Source: Kraft General Foods Corp.

Coconut Ice Cube Center

			% by Wt.
Gem Medium Coconut	33	lb	31.69
Invert sugar	6.5	lb	6.24
Sugar	16.5	lb	15.85
Corn syrup	33	lb	31.70
Creamed Coconut	10	lb	9.60
Vanilla extract	2	oz	0.12
Nougat cream (q.v.)	5	lb	4.80
			Total 100.00

Procedure:

Place **Gem Medium Coconut** and melted invert sugar in mixer and mix for 15 min.

Cook sugar, corn syrup, and salt to 234 F.

Add **Creamed Coconut** and vanilla. Stir well. Add to tenderized **Gem Medium Coconut**. Add nougat cream and mix well. Let cool and form into desired shapes.

NOUGAT CREAM:

Sugar	20	lb
Corn syrup	40	lb
Egg albumen	1	lb
Cold water	2	lb

Procedure:

Cook sugar to 232 F. Add corn syrup and place in beater. Add egg albumen which has been soaked in water. Beat to a dry beat.

Source: Kraft General Foods Corp.

Coconut Ice Tea-n-Lemon Bitsyn® Bar

			% by Wt.
White sugar	6	lb	44.04
Honey	1.5	lb	11.02
Water	2	lb	14.68
Fondant 4-1	1.25	lb	9.17
Coconut medium desiccated*	1.25	lb	9.17
Bitsyn® PC 1761*	8	oz	3.67
Tea-n-Lemon	6	oz	2.75
Bitsyn® PC 1766	12	oz	5.50
			Total 100.00

Procedure:

In copper kettle cook sugar, honey, and water to 239 F. Cool to 200 F. Start a grain by agitating with wooden spatula. When grain is well started, add

* Combine

fondant, mixing thoroughly. Add **Tea-n-Lemon**, then the combined coconut and small **Bitysn®**, mixing to dissolve any lumps. Next put in the **Bitsyn® PC 1766.** Pour onto silicone paper placed on a warm slab and spread between 3/4 in. bars. When grained, cut into bars. Coat in milk chocolate.

Source: Pillsbury Co.

Coconut Kisses

			% by Wt.
Amerfond fondant sugar	50	lb	60.45
Water	7.25	lb	8.77
Salt	6	oz	0.36
Frappé	15	lb	18.14
Macaroon coconut	10	lb	12.09
Invertase	2.5	oz	0.19
Flavoring			to suit
			Total 100.00

Procedure:

Place fondant sugar in remelt kettle or double boiler. Add water and salt and stir, with heat, until smooth. Stir in remaining ingredients and heat to 160 F.

Spoon out onto oiled or silicone paper, or extrude or deposit.

These may be partially coated with chocolate if desired.

Source: Amstar Co.

Coconut Kraut

			% by Wt.
Gem Medium Coconut	15	lb	23.62
Nulomoline (invert sugar)	5	lb	7.87
Water	2	lb	3.15
Corn syrup	15	lb	23.62
Brown sugar	2.5	lb	3.94
White sugar	10.5	lb	16.54
Condensed milk	8	lb	12.60
Creamed Coconut	1.5	lb	2.36
Fondant 80-20	4	lb	6.30
			Total 100.00

Procedure:

Place **Gem Medium Coconut**, melted **Nulomoline**, and water in dough mixer and mix well.

Place sugar, brown sugar, corn syrup, and water to dissolve sugar in steam kettle. Bring to boil and slowly add evaporated milk. Cook to 244–246 F. Add **Creamed Coconut** and stir well.

SUGGESTIONS:

If darker color is desired, add caramel color.

Source: Kraft General Foods Corp.

Coconut Mallo Cream

					% by Wt.
Sugar	24	lb	7	oz	23.43
Corn syrup	11	lb	8	oz	11.50
Water	9	lb	4	oz	9.25
Corn starch	2	lb	5	oz	2.31
Water	9	lb	4	oz	9.25
Corn syrup	23	lb	6	oz	23.38
Egg albumen	1	lb	4	oz	1.25
Water	2	lb	14	oz	2.88
Creamed Coconut	16	lb	12	oz	16.75
					Total 100.00

Procedure:

Place sugar, first amount of water, and corn syrup in steam-jacketed kettle and bring to a boil.

Dissolve corn starch in water and add to boiling syrup. Cook to 236 F and shut off steam.

Add second portion of corn syrup and water to cooked mixture and stir well. Pour boiling syrup in mixer and beat until temperature is about 160 F.

Slowly add egg albumen to syrup and whip until fluffy.

Lastly add melted **Creamed Coconut** and mix well.

Source: Kraft General Foods Corp.

Coconut Nougat

			% by Wt.
Gem Medium Coconut	6	lb	9.44
Invert sugar	9	oz	0.88
Nougat cream	18	lb	28.32
Corn syrup	12	lb	18.88
Sugar	18	lb	28.32
Egg albumen	1	lb	1.57
Water	2	lb	3.15
Creamed Coconut	6	lb	9.44
			Total 100.00

Procedure:

Place **Gem Medium Coconut** and melted invert sugar in mixer. Blend well.

Place nougat cream in high-speed beter. Cook corn syrup and sugar to 280 F. Add slowly to nougat cream in beater.

Add egg albumen soaked in water and beat until stiff.

Melt the **Creamed Coconut** in a large clean kettle. Pour nougat out of beater into kettle and fold in **Creamed Coconut** with a paddle. Lastly add freshened coconut.

NOUGAT CREAM:

Sugar	20	lb
Corn syrup	40	lb
Egg albumen	1	lb
Cold water	2	lb

Procedure:

Cook sugar to 232 F. Add corn syrup and place in beater. Add egg albumen which has been soaked in water.

Source: Kraft General Foods Corp.

Coconut Orange Slices

			% by Wt.
Gem Medium Coconut	2	lb	8.74
Invert sugar	8	oz	2.19
Water	4	oz	1.09
Fondant 80-20	10	lb	43.72
Creamed Coconut	1	lb	4.37
Sugar	6	lb	26.23
Corn syrup	1	lb	4.37
Frappé (q.v.)	2	lb	8.74
Convertit (invertase)	2	oz	0.55
Orange color and flavor			to suit
			Total 100.00

Procedure:

Place **Gem Medium Coconut**, melted invert sugar, and water in mixer. Blend well.

Place fondant, **Creamed Coconut**, and tenderized **Gem Medium Coconut** in remelt kettle.

Combine sugar, corn syrup, and water to dissolve sugar in kettle. Cook to 244 F. Pour syrup over fondant and **Creamed Coconut**.

Lastly, add frappé, **Convertit**, color, and flavor. Mix well. Cast in starch. When dry, shake out and enrobe in chocolate.

FRAPPÉ:

Sugar	20	lb
Corn syrup	40	lb
Egg albumen	1	lb
Cold water	2	lb

Procedure:

Cook sugar to 232 F. Add the corn syrup and place in beater. Add the egg albumen which has been dissolved in water. Beat to a dry beat.

Source: Kraft General Foods Corp.

Coconut Pineapple Fudge

			% by Wt.
Sugar	25	lb	34.68
Corn syrup	5	lb	6.94
Crushed pineapple (drained slightly)	8	lb	11.10
Creamed Coconut	2	lb	2.78
Fondant 80-20 (q.v.)	12	lb	16.65
Pineapple flavor	1	oz	0.09
Frappé (q.v.)	10	lb	13.88
Gem Long Thread Coconut	10	lb	13.88
			Total 100.00

Procedure:

Place sugar, corn syrup, crushed pineapple, and water to dissolve sugar in steam kettle. Cook to 242 F.

Add **Creamed Coconut**, fondant, pineapple flavor, and frappé. Stir until batch has a creamy texture.

Add **Gem Long Thread Coconut** and mix well. Pour onto a slightly greased slab to about 1 in. thick. Let set until firm. Cut to desired shape and size.

FONDANT 80-20:

Sugar	80	lb
Corn syrup	20	lb
Water (to dissolve sugar)		

Procedure:

Cook to 244 F.

FRAPPÉ:

Sugar	20	lb
Corn syrup	40	lb
Egg albumen	1	lb
Cold water	2	lb

Procedure:

Cook sugar to 232 F. Add corn syrup and place in beater. Add egg albumen which has been dissolved in water. Beat to a dry beat.

Source: Kraft General Foods Corp.

Coconut Potatoes

			% by Wt.
Plastic coconut	15	lb	12.21
Nulomoline (invert sugar)	15	lb	12.21
Corn syrup	15	lb	12.21
Water	2.5	lb	2.03
Sucrovert (liquid invertase)	2	oz	0.10
Icing sugar	75	lb	61.04
Salt	4	oz	0.20
Vanilla flavor			to suit
			Total 100.00

Procedure:

Heat the **Nulomoline**, corn syrup, plastic coconut and water to approximately 120 F. Add the salt, **Sucrovert**, and flavor. Mix well. Add additional icing sugar or water to obtain the desired consistency. When rolling or forming the batch, dust well and roll with the following dusting material:

CINNAMON DUSTING MATERIAL:

Powdered cinnamon	1	oz
Icing sugar, plus sufficient brown color to simulate potato skin color	20	oz

Procedure:

Mix thoroughly, then pass through a fine sieve to remove lumps.

FRUIT FORMS:

The above plastic coconut paste may also be formed to simulate apples, oranges, strawberries, or carrots, allowed to dry on the surface, then glazed with Candy-Lac.

Source: Crompton & Knowles

Coconut Pralines

			% by Wt.
Gem Coarse Coconut	1	lb	7.83
Invert sugar	3	oz	1.47
Water	1	oz	0.49

Brown sugar	4	lb	31.34
White sugar	3	lb	23.50
Sweetose (corn syrup)	1	lb	7.83
Water	3	lb	23.50
Salt	0.25	oz	0.12
Creamed Coconut	8	oz	3.92
			Total 100.00

Procedure:

Place **Gem Coarse Coconut** in mixer with melted invert sugar and water. Mix well.

Combine brown sugar, white sugar, **Sweetose**, salt and water in kettle. Cook to 236 F.

Add **Creamed Coconut** and stir until dissolved. Cook to 130–140 F then beat until batch starts to grain. Lastly add tenderized coconut. Mix well and cast in cups. When set, wrap in waxed paper or cellophane.

Source: Kraft General Foods Corp.

Coconut Raspberry Tropical Jells

			% by Wt.
Water	20.75	lb	24.43
Modified pectin*	1	lb	1.18
Sugar	25	lb	29.43
Corn syrup	35	lb	41.21
M.P. Buffer*	3	oz	0.22
Creamed Coconut	3	lb	3.53
Raspberry flavor and color			to suit
			Total 100.00

Procedure:

Dry mix modified pectin with part of the sugar and stir into water. Boil until thoroughly dissolved. Now add the balance of sugar, then the corn syrup. When the batch boils add the **M.P. Buffer**, mixed with 4 or 5 oz of water. After the batch has boiled for a few seconds, add the **Creamed Coconut**. Cook the

* Obtained from T.H. Augermeir & Co., 245 7th Ave., New York, NY.

batch to 226 F; turn off steam. Allow batch to stop boiling. Add raspberry flavor and color. If desired add some frappé to lighten batch cast.

SUGGESTIONS:

If batch is to be poured on slab, add **M.P. Buffer** just before sugar and corn syrup are added. When jelly has set, cut into squares and roll in **Gem Medium Coconut**.

Source: Kraft General Foods Corp.

Coconut Rough Tops

(Formed on Rolled Cream Center Machine)

			% by Wt.
PART 1:			
Medium desiccated coconut	24	lb	15.32
Macaroon desiccated coconut	20	lb	12.76
Nulomoline (invert sugar)	16	lb	10.21
PART 2:			
Granulated sugar	30	lb	19.15
Corn syrup	12	lb	7.66
Water	8	lb	5.11
PART 3:			
Casting fondant	40	lb	25.53
Frappé No. 1 or 2 (q.v.)	6	lb	3.82
Convertit (invertase)	3	oz	0.12
Salt	8	oz	0.32
Vanilla flavor (white portion)			to suit
Grape flavor (brown portion)			to suit
			Total 100.00

Procedure:

PART 1: Mix altogether well without heating.

PART 2: Cook altogether to 244 F, then add the freshened coconut (Part 1), and mix well. Immediately add Part 3.

Mix Part 3 thoroughly and set aside until the batch becomes moderately plastic. Then color half of the batch brown. Place alternate layers of the white and brown batch into the hopper of the rolled cream center machine and extrude any shape die.

The formed candies may then be placed into paper cups or may be partly coated on the bottom with chocolate.

FRAPPÉ NO. 1:

Corn syrup	50 lb
Invert sugar	50 lb
Albumen	1 lb
Water	2 lb

Procedure:

Cook corn syrup to 245 F (sea level), shut off the heat and add invert sugar and stir until it is melted. Place the batch into a marshmallow beater, start the beater, then gradually add albumen which has been previously dissolved in water and beat the batch until light.

FRAPPÉ NO. 2:

Corn syrup	25 lb
Nulomoline (invert sugar)	25 lb
Albumen (dissolved in the water)	1 lb
Water (cold)	2 lb

Procedure:

Boil the corn syrup to 245 F, add the **Nulomoline** and mix until it is completely dissolved. Turn off the heat, place the batch into a beating machine, start the beater, then gradually add the albumen solution, and beat the batch until it is quite stiff. Remove the frappé from the beater and place into a suitable container for future use.

Source: Crompton & Knowles

Coconut Short Nougat

		% by Wt.
Nougat cream (q.v.)	18 lb	32.14
Corn syrup	12 lb	21.43

Sugar	18	lb	32.14
Egg albumen	1	lb	1.79
Water	2	lb	3.57
Creamed Coconut	5	lb	8.93
			Total 100.00

Procedure:

Place nougat cream in high-speed beater.

Combine sugar, corn syrup, and water to dissolve sugar. Cook to 280 F. Add slowly to nougat cream in beater.

Add egg albumen soaked in water; beat until stiff.

Melt **Creamed Coconut** in large, clean kettle. Pour out of beater and fold in **Creamed Coconut** with a paddle. Pour onto a paper-lined cold-water slab to a 1 in. thickness. Let set until nougat grains. Cut to desired shapes and sizes.

Before cutting this piece, a thin layer of melted chocolate is spread on nougat and then covered with **Golden Toasted Rice Cut Coconut**. Coconut is patted or rolled into chocolate. When chocolate has dried, cut and enrobe in chocolate.

NOUGAT CREAM:

Sugar	20	lb
Corn syrup	40	lb
Egg albumen	1	lb
Cold water	2	lb

Procedure:

Cook the sugar to 232 F. Add corn syrup and place in beater. Add egg albumen that has been soaked in water. Beat to a dry texture.

Source: Kraft General Foods Corp.

Coconut Snowdrift Fudge

			% by Wt.
Gem Medium Coconut	15	lb	13.34
Nulomoline (invert sugar)	5	lb	4.44
Water	4	lb	3.55
Corn syrup	15	lb	13.33
Sugar	30	lb	26.66

Creamed Coconut	5	lb	4.44
Fondant 70-30	30	lb	26.66
Salt	4	oz	0.22
Vanilla	4	oz	0.22
Nougat cream	8	lb	7.11
Invertase	$^1/_2$	oz	0.03
			Total 100.00

Procedure:

Place in mixer and blend together **Gem Medium Coconut** and melted **Nulomoline.**

Place corn syrup, sugar, **Creamed Coconut** together in a clean kettle with water to dissolve sugar and cook to 244 F.

Add fondant, salt, and vanilla to bob syrup. Then add tenderized **Gem Medium Coconut.** Next add nougat cream and finally add invertase. Mix well. Spread on table lined with waxed paper. When set, cut to desired size.

Source: Kraft General Foods Corp.

Coconut Soft Center

No. 1

			% by Wt.
Gem Medium Coconut	70	lb	37.49
Nulomoline (invert sugar)	30	lb	16.06
Water	11	lb	5.89
Corn syrup	70	lb	37.48
Creamed Coconut	5	lb	2.68
Salt	7	oz	0.23
Vanilla flavor	5	oz	0.17
			Total 100.00

Procedure:

Heat the **Nulomoline** and water to 170–180 F. Add **Gem Medium Coconut.** Mix well.

Heat corn syrup (do not boil). Add **Creamed Coconut**, mix until dissolved. Now pour batch over freshened coconut, add salt and vanilla flavor. Mix well, cool, and form into desired shapes.

Source: Kraft General Foods Corp.

No. 2

			% by Wt.
Corn syrup	20	lb	15.93
Granulated sugar	50	lb	39.83
Gem Macaroon Coconut	30	lb	23.89
Nulomoline (invert sugar)	10	lb	7.96
Water	1	lb	0.80
Creamed Coconut	4	lb	3.19
Hard butter (84 F)	5	lb	3.98
Frappé	5	lb	3.98
Vanilla extract	3	oz	0.15
Vanilla sugar (q.v.)	2	oz	0.09
Salt	4	oz	0.20
			Total 100.00

Procedure:

Add water to dissolve sugar, combine with corn syrup and cook to 228 F.

Mix **Gem Macaroon Coconut, Nulomoline**, and water and allow to stand for 15 min.

Add **Creamed Coconut** and hard butter to bob when cook is finished. Then add freshened coconut.

Add frappé to batch to lighten color.

Add vanilla extract, vanilla sugar, and salt to finished batch and then cool to 90 F for depositing.

SUGGESTIONS:

Be sure chocolate shell is thoroughly chilled before depositing center.

VANILLA SUGAR:

Sugar (6X)	98	oz
Monsanto Ethavan	1.25	oz
Monsanto Coumarin	0.75	oz

Source: Kraft General Foods Corp.

Coconut Tea Biscuits

			% by Wt.
Gem Medium Coconut	5	lb	35.64
Nulomoline (invert sugar)	8	oz	3.56
Sugar	4	lb	28.51
Corn sugar	4	lb	28.51
Salt	1/4	oz	0.11
Creamed Coconut	8	oz	3.56
Vanilla extract	1/4	oz	0.11
			Total 100.00

Procedure:

Place **Gem Medium Coconut** in mixer bowl and pour melted **Nulomoline** over. Mix at low speed for 15 min.

Place sugar and corn sugar in clean kettle with water to dissolve sugar. Cook to 238 F.

Now add salt and **Creamed Coconut**. Stir until **Creamed Coconut** is well dissolved in syrup. Then add vanilla extract. Pour this syrup over the coconut in the mixer and mix at low speed for 1–2 min. Pour on slab, spread out, roll down and cut out as desired and run through hot oven until tops are toasted.

SUGGESTIONS:

If tea biscuit is too soft, raise temperature to 240 F.

Source: Kraft General Foods Corp.

Coconut Wafer Fill

			% by Wt.
Creamed Coconut	25	lb	25.00
Shortening (70 F)	20	lb	20.00
Gem Macaroon Coconut	5	lb	5.00
4X powdered sugar	49.4	lb	49.40
Powdered salt	8	oz	0.50
Vanilla sugar	1.5	oz	0.10
			Total 100.00

Procedure:

Cream together **Creamed Coconut** and shortening for a uniform coconut flavor.

Add remainder of ingredients slowly to above mixture at medium speed in mixer. When all materials are in, whip until light at high speed.

SUGGESTIONS:

If desired, substitute half of 4X sugar with dextrose. In summer, substitute 110 F vegetable hard butter for part or all of 70 F shortening.

Source: Kraft General Foods Corp.

Cream Coconut Croquettes, Cakes, Etc.

(Short in Texture)

		% by Wt.
PART 1:		
Medium shred coconut	5 lb	15.12
Nulomoline (invert sugar)	1 lb	3.02
Water	8 oz	1.51
PART 2:		
Sugar	7 lb	21.17
Corn syrup	1 lb	3.02
Water	2 lb	6.05
PART 3:		
Fondant (q.v.)	15 lb	45.39
Frappé No. 1 (q.v.)	1.5 lb	4.53
Convertit (invertase)	1 oz	0.19
Flavor and color		to suit
		Total 100.00

Procedure:

PART 1: Mix the water with the **Nulomoline**, add the coconut, mix well, and set aside.

PART 2: Boil altogether to 250–254 F.

Add the freshened coconut (Part 1) to Part 2. Then add fondant.

Mix the batch until the fondant is melted, then add Frappé No. 1, **Convertit**, flavor and color as desired.

Spoon the batch out on waxed paper into croquettes, cakes, bars, etc.

FONDANT:

Sugar	40 lb
Nulomoline (invert sugar)	5 lb
Corn syrup	5 lb
Water (to dissolve sugar)	

Procedure:

Boil altogether to 244–246 F, then cool and beat into fondant.

FRAPPÉ NO. 1:

Corn syrup	50 lb
Invert sugar	50 lb
Albumen	1 lb
Water	2 lb

Procedure:

Cook corn syrup to 245 F (sea level), shut off the heat and add invert sugar and stir until it is melted. Place the batch into a marshmallow beater, start the beater, then gradually add albumen which has been previously dissolved in water and beat the batch until light.

Source: Crompton & Knowles

Creamed Coconut Caramels

		% by Wt.
Sugar	20 lb	14.79
Sweetened condensed milk	40 lb	29.57
Corn syrup	60 lb	44.36
Creamed Coconut	15 lb	11.09
Salt	4 oz	0.19
		Total 100.00

Procedure:

Place sugar, corn syrup, and milk in kettle. Cook to a medium boil.

Add **Creamed Coconut** and salt. Stir well. Pour on greased slab. Spread with coconut center and roll into pinwheels. Cut and wrap on kiss machine.

Source: Kraft General Foods Corp.

Creamed Coconut Toffee

			% by Wt.
Sweetened condensed milk	50	lb	43.38
Corn syrup	50	lb	43.38
Creamed Coconut	15	lb	13.02
Salt	4	oz	0.22
			Total 100.00

Procedure:

Combine milk and corn syrup in agitator-equipped steam kettle. Cook to a medium boil.

Add salt and **Creamed Coconut**. Mix thoroughly and pour on slab to cool. Cut as desired.

SUGGESTIONS:

May be used as a jacket around coconut center.

Source: Kraft General Foods Corp.

French Coconut Creams

				% by Wt.
Evaporated milk	2 lb	10	oz	10.51
Sugar (4X or 6X)	2 lb			8.00
Gem Medium Coconut	4 lb	7	oz	17.76
Cinnamon		1/3	oz	0.08
Fondant 90-10	6 lb	10	oz	26.54
Dark chocolate	4 lb			16.01
Milk chocolate	2 lb	10	oz	10.51
Creamed Coconut	2 lb	10	oz	10.51
Convertit (invertase)		1/3	oz	0.08
				Total 100.00

Procedure:

Place evaporated milk, sugar, **Gem Medium Coconut**, and cinnamon in Hobart mixer. Blend well at low speed.

Melt fondant and add to the above ingredients.

Melt chocolate and **Creamed Coconut** together and add to mixture. Lastly add **Convertit** and mix at second speed for 1 min.

Let candy set for 2 h before cutting.

Source: Kraft General Foods Corp.

Fruit–Nut–Coconut Bars

			% by Wt.
PART 1:			
Desiccated coconut, medium shred	15	lb	10.79
Water	6	lb	4.32
Nulomoline (invert sugar)	6	lb	4.32
PART 2:			
Sugar	50	lb	35.97
Corn syrup	10	lb	7.19
Nulomoline (invert sugar)	10	lb	7.19
Water (to dissolve sugar)			
PART 3:			
Special Frappé (q.v.)	25	lb	17.98
Sliced Brazil nuts	6	lb	4.32
Preserved pineapple pieces	8	lb	5.76
Preserved cherry pieces	3	lb	2.16
		Total	100.00

Procedure:

PART 1: Mix the water with the **Nulomoline**, add the coconut, mix well, and set aside for 10 or 15 min.

PART 2: Boil altogether to 275–280 F.

Pour Part 2 at once on top of the freshened coconut (Part 1), then add Special Frappé. Mix well and add nuts and fruit pieces.

Mix well and spread the batch out on waxed or oiled paper to set. When cold, cut into bars of the desired size which may later be coated with chocolate, icing, or jelly.

SPECIAL FRAPPÉ:

Corn syrup	50	lb
Nulomoline (invert sugar)	50	lb
Albumen (dissolved in the water)	3	lb
Cold water	6	lb
Tartaric acid	1	oz

Procedure:

Boil the corn syrup to 245 F, add the **Nulomoline** and mix until melted. Place the batch at once into a marshmallow beater, start the beater, then gradually add the albumen solution and tartaric acid; beat until light.

Source: Crompton & Knowles

Gem Slice Confection

			% by Wt.
Gem Slice Coconut	56 lb 9	oz	96.18
Corn syrup	12	oz	1.28
Sugar	12	oz	1.28
Water	6	oz	0.63
Salt	6	oz	0.63
			Total 100.00

Procedure:

Place corn syrup, sugar, water, and salt in kettle. Heat until dissolved. Pour syrup over **Gem Slice Coconut** and mix well. Toast at 275–300 F until desired color and glaze are reached.

Source: Kraft General Foods Corp.

Grained Coconut Tips

(Friend Machine or Racine Depositor)

			% by Wt.
PART 1:			
Medium desiccated coconut	20	lb	10.05
Macaroon desiccated coconut	20	lb	10.05
Nulomoline (invert sugar)	10	lb	5.03
Salt	1	lb	0.50
PART 2:			
Granulated sugar	50	lb	25.12
Corn syrup	20	lb	10.05
Water	8	lb	4.02
PART 3:			
Basic Fondant	60	lb	30.15
Frappé No. 1 (q.v.)	10	lb	5.03
Vanilla or other flavor and color			to suit
			Total 100.00

Procedure:

PART 1: Mix together, cool, and set aside for 15 min or longer.

PART 2: Cook the sugar, corn syrup, and water to 250–254 F.

Add Part 1 to Part 2, mixing well, then add Part 3.

Mix thoroughly and allow the batch to cool until it is of moderate plasticity, then place it into the hopper of the Friend or Racine depositor and deposit onto heavy waxed paper.

FRAPPÉ NO. 1:		
Corn syrup	50	lb
Invert sugar	50	lb
Albumen	1	lb
Water	2	lb

Procedure:

Cook corn syrup to 245 F (sea level), shut off the heat and add invert sugar and stir until it is melted. Place the batch into a marshmallow beater, start the beater, then gradually add albumen which has been previously dissolved in water and beat the batch until light.

COCOA FLAVOR:

For cocoa-flavored batch, cook Part 2 to approximately 246 F, and after Part 1 has been mixed with Part 2, add approximately 8 lb of dutched process cocoa powder, mixing well, then add Part 3, plus vanilla flavor, mix well and deposit.

Source: Crompton & Knowles

Honeycomb Creamed Coconut Chips

			% by Wt.
Corn syrup	15	lb	31.58
Granulated sugar	25	lb	52.63
Creamed Coconut	5	lb	10.53
Sugar (4X)	2.5	lb	5.26
			Total 100.00

Procedure:

Place ingredients together with water to dissolve. Cook to 310 F. Set aside a small amount of clear candy for jacket. Pull remainder of batch up light and spread out and place warm **Creamed Coconut** which has been stiffened with powdered sugar in center of jacket. Fold jacket to cover coconut. Weave eight times. Spin out and cut in chip shape.

SUGGESTIONS:

For variety add butter to hard candy batch and make butterscotch. Add molasses to cooked hard candy before pulling light.

Source: Kraft General Foods Corp.

Iced Banana Coconut Bonbons

			% by Wt.
Nulomoline (invert sugar)	30	lb	15.04
Water	10	lb	5.01
Corn syrup	70	lb	35.10
Desiccated macaroon coconut	75	lb	37.61
Salt	7	oz	0.22

Nulco MeloBan (powdered ripe banana)	4	lb	2.00
Banana color			to suit
Ground raisins	10	lb	5.02
			Total 100.00

Procedure:

Heat the **Nulomoline** and water to approximately 150 F, add the coconut, mix well, then sift in the **Nulco MeloBan.** Add the salt, color, corn syrup, and ground raisins and mix well.

Form the batch into balls or ovals and coat with icing, fondant, or chocolate.

Source: Crompton & Knowles

Maple Coconut Creams

			% by Wt.
PART 1:			
Gem Macaroon Coconut	4.25	lb	4.29
Invert sugar	1	lb	1.01
Water hot	8	oz	0.51
Salt	2.5	oz	0.16
PART 2:			
Water	14	lb	14.14
Sugar	61	lb	61.62
Invert sugar	7	lb	7.07
Invertase	1.5	oz	0.09
Tenderized coconut (Part 1)			
Creamed Coconut	2	lb	2.02
Maple flavor and color			to suit
Nougat cream	9	lb	9.09
			Total 100.00

Procedure:

PART 1: Add water, salt, invert sugar, and **Gem Macaroon Coconut** and mix thoroughly. Then spread this over open-type fondant beater.

PART 2: Place sugar, balance of invert sugar and water in steam kettle. After batch starts to boil, wash down all sugar crystals that cling to kettle.

Cook batch to 244–246 F. Pour cooked batch over coconut preparation in fondant beater and immediately sprinkle the surface of batch with not more than 2 oz of water. Cook batch to 100 F and add **Creamed Coconut** that has been cut into small pieces. Start beater. Add invertase and flavor. Continue beating until batch shows evidence of graining. Add frappé beating until batch grains. Allow batch to set 5–10 min, knead until smooth. Form centers on Friend machine.

NOUGAT CREAM:

Sugar	20 lb
Corn syrup	40 lb
Egg albumen	1 lb
Cold water	2 lb

Procedure:

Sugar is cooked to 232 F. Add corn syrup and place in beater. Add egg albumen which has been dissolved in the water and beat dry.

Source: Kraft General Foods Corp.

MeloBan Fruit Coconut Bars

		% by Wt.
Sugar	20 lb	20.73
Corn syrup	20 lb	20.73
Nulomoline (invert sugar)	12 lb	12.44
Water (to dissolve the sugar)		
Sliced Brazil nuts (colored green)	4 lb	4.14
Desiccated macaroon coconut (mixed with 4 lb water)	25 lb	25.90
Frappé No. 1 (q.v.)	3 lb	3.10
Nulco MeloBan (powdered ripe banana)	2.5 lb	2.60
Preserved cherries, figs, pineapple, lemon, orange peel	10 lb	10.36
Banana color		to suit
		Total 100.00

Procedure:

Mix the coconut with 4 lb of water.

Place the corn syrup, sugar, and **Nulomoline** into a kettle with sufficient water to dissolve the sugar; boil the batch to 245 F. Add the freshened coconut, mix well and continue to cook and stir the batch to a firm paste. Now add the frappé, then sift in **Nulco MeloBan**, add banana color and Brazil nuts, and mix thoroughly.

Spread half of the batch on an oiled and dusted slab about $^3/_4$ in. high. Spread the preserved fruits over the batch on the slab, then immediately cover the preserved fruit with the balance of the batch and roll it down to the required height, and when cold, cut into bars, oblongs, squares, etc. Roll in fruit powdered sugar.

FRAPPÉ NO. 1:	
Corn syrup	50 lb
Invert sugar	50 lb
Albumen	1 lb
Water	2 lb

Procedure:

Cook corn syrup to 245 F (sea level), shut off the heat and add invert sugar and stir until it is melted. Place the batch into a marshmallow beater, start the beater, then gradually add albumen which has been previously dissolved in water and beat the batch until light.

Source: Crompton & Knowles

MeloBan Coconut Bar

		% by Wt.
PART 1:		
Desiccated coconut	65 lb	27.98
Nulomoline (invert sugar)	20 lb	8.61
Corn syrup	20 lb	8.61
Water	10 lb	4.30
PART 2:		
Granulated sugar	50 lb	21.52
Corn syrup	30 lb	12.91
Water	12 lb	5.16

PART 3:

Basic Fondant (q.v.)	10	lb	4.30
Salt	4	oz	0.11
Convertit (invertase)	2	oz	0.05
Frappé No. 1 (q.v.)	10	lb	4.30
Nulco MeloBan (powdered ripe banana)	5	lb	2.15
Yellow color			to suit
			Total 100.00

Procedure:

PART 1: Heat the corn syrup, **Nulomoline**, and water to approximately 150 F, add the coconut and mix thoroughly, then set aside for at least $^1/_2$ h so as to give the coconut time to absorb the moisture.

PART 2: Boil the sugar, corn syrup, and water to 250 F.

Pour Part 2 at once onto the freshened coconut (Part 1) and mix thoroughly. Then add Part 3 (except **MeloBan** and color) and mix well. Then sift in **Nulco MeloBan**, add yellow color, mix thoroughly and spread batch out to set.

BASIC FONDANT:

Granulated sugar	80	lb
Corn syrup	20	lb
Invert sugar	10	lb
Water		

Procedure:

Boil granulated sugar, corn syrup, and invert sugar, plus sufficient water to dissolve the sugar, to 242 F (sea level), then cool and cream into fondant.

FRAPPÉ NO. 1:

Corn syrup	50	lb
Nulomoline (invert sugar)	50	lb
Albumen (dissolved in the water)	1	lb
Water	2	lb

Procedure:

Boil the corn syrup to 245 F, shut off steam, add the **Nulomoline** and stir until it is melted. Place the batch into a marshmallow beater, start the beater and gradually add the albumen solution and beat until light.

Source: Crompton & Knowles

Molasses Coconut Chews

			% by Wt.
Corn syrup	45	lb	22.64
Granulated sugar	15	lb	7.55
Molasses (**Grandma's**)	45	lb	22.64
Desiccated coconut (1/2 macaroon, 1/2 medium or long shred)	60	lb	30.20
Gelatin (200 bloom, mixed with water)	12	oz	0.38
Hot water	1.5	lb	0.75
Sweetened condensed whole milk	30	lb	15.09
Salt	1.5	lb	0.75
			Total 100.00

Procedure:

Place the corn syrup, sugar, molasses, condensed milk, and salt into a caramel cooking kettle, heat the batch until it boils, then add the desiccated coconut and continue to stir and cook the batch to a medium firm ball.

Add the gelatin–water combination, mix thoroughly and spread the batch on an oiled slab lightly sprinkled with fine granulated sugar. Roll the batch down to form a sheet approximately 1 in. thick, sprinkling the surface with granulated sugar.

When the batch is firm, it may be cut into slices, oblongs, or squares and rolled in fine granulated sugar.

Brazil nuts, peanuts, almonds, or walnuts may be added to the above batch if desired.

Source: Crompton & Knowles

Molasses Coconut Rolls

(Rolled in Toasted Coconut)

No. 1

			% by Wt.
Molasses (**Grandma's**)	40	lb	24.21
Corn syrup	60	lb	36.31
Desiccated macaroon coconut	65	lb	39.33
Salt	4	oz	0.15
			Total 100.00

Procedure:

Heat the molasses to approximately 150 F. Add the coconut and mix thoroughly. If possible permit this to set for at least 1/2 h so that the coconut will be softened. Then add the corn syrup and salt; stir and cook batch to a medium soft ball. Spread batch out to cool and roll into shapes of the desired size or cut into squares.

Moisten the candies with water icing (q.v.) or simple syrup and roll in toasted coconut, or dip in bonbon cream.

WATER ICING:		
Powdered sugar (XXXX)	25	lb
Cold water	25	lb
Powdered sugar (XXXX)	50	lb

Procedure:

Place 25 lb of sugar into a marshmallow beater. Gradually add the cold water and beat well; then add 50 lb more of sugar and beat well.

Set this icing aside until ready for use, at which time it may be thinned down to the required consistency.

Source: Crompton & Knowles

No. 2

			% by Wt.
Molasses	40	lb	24.21
Desiccated macaroon coconut	65	lb	39.33
Corn syrup	60	lb	36.31
Salt	4	oz	0.15
			Total 100.00

Procedure:

Heat the molasses to approximately 180 F, add the coconut and mix thoroughly. If possible, set the batch aside for at least 1/2 h to allow the coconut to soften. Now add the corn syrup and salt. Stir and cook the batch to a medium-soft ball.

Spread the batch out to cool and roll into shapes of the desired size or cut into squares. Moisten the candies with icing or simple syrup and roll in toasted coconut.

Source: Crompton & Knowles

Molasses Coconut Slices

		% by Wt.
PART 1:		
Granulated sugar	20 lb	18.52
Evaporated milk	4 lb	3.70
Corn syrup	24 lb	22.23
Molasses (**Grandma's**)	8 lb	7.41
Water	4 lb	3.70
Desiccated macaroon coconut	24 lb	22.23
PART 2:		
Fondant (q.v.)	12 lb	11.11
Frappé No. 1 (q.v.)	4 lb	3.70
Broken Brazil nuts	8 lb	7.40
Flavor and salt		to suit
		Total 100.00

Procedure:

PART 1: Place altogether, except for the coconut, into a caramel cooking kettle. Bring to a boil, then add the coconut. Continue to stir and cook to a medium firm ball.

Then add Part 2. Mix thoroughly and spread the batch out on an oiled slab lightly dusted with finely granulated sugar. Sprinkle granulated sugar over the batch and roll it down to the required height (about 1 in. thick). When the batch is cold, it may be cut into slices, squares, or oblongs, then rolled in fine granulated sugar, then packed as is, or into paper cups, or the candies may be coated with chocolate icing or colored confectioners coating.

FONDANT:

Sugar	75 lb
Corn syrup	25 lb

Procedure:

Cook to 245 F.

FRAPPÉ NO. 1:

Corn syrup	50 lb
Invert sugar	50 lb
Albumen	1 lb
Water	2 lb

Procedure:

Cook corn syrup to 245 F (sea level), shut off the heat and add invert sugar and stir until it is melted. Place the batch into a marshmallow beater, start the beater, then gradually add albumen which has been previously dissolved in water and beat the batch until light.

Source: Crompton & Knowles

Molasses Coconut Stripes

			% by Wt.
Gem Macaroon Coconut	30	lb	23.62
Nulomoline (invert sugar)	10	lb	7.87
Corn syrup	55	lb	43.32
Sugar	20	lb	15.75
Molasses	10	lb	7.87
Creamed Coconut	2	lb	1.57
			Total 100.00

Procedure:

Place **Gem Macaroon Coconut** and melted **Nulomoline** in dough mixer and mix well.

Cook corn syrup, sugar, and molasses to 246 F. Add **Creamed Coconut** and stir well. Now add tenderized **Gem Macaroon Coconut** to bob and mix well. Pour on slab that has been covered with granulated sugar. Roll down to desired height. When cool, cut into strips and roll in granulated sugar.

Source: Kraft General Foods Corp.

Orange Coconut Creams

			% by Wt.
Sugar	45	lb	61.57
Invert sugar	5	lb	6.84
Frappé (q.v.)	3	lb	4.10
Creamed Coconut	5	lb	6.84
Gem Macaroon Coconut	5	lb	6.84

Whole oranges (ground)	5 lb	6.84
Convertit (invertase)	5 lb	6.84
Orange oil and color	1.5 oz	0.13
		Total 100.00

Procedure:

Add water to dissolve sugar. Bring to boil. Wash down sugar crystals from side of kettle. Cook to 244 F. Pour batch on open-type cream beater.

Cool batch to 100 F and start beater. When batch begins to grain, add frappé, **Gem Macaroon Coconut, Creamed Coconut**, oranges, color, flavor, and **Convertit.** Let batch set until firm. When set, deposit or roll by hand. Dip in chocolate.

FRAPPÉ:

Sugar	20 lb
Corn syrup	40 lb
Egg albumen (dissolved in water)	1 lb
Cold water	2 lb

Procedure:

Cook sugar to 232 F. Add corn syrup and place in beater. Add egg albumen which has been dissolved in water. Beat to a dry beat.

Source: Kraft General Foods Corp.

Orange Coconut Jellies

		% by Wt.
Gem Medium Coconut	15 lb	16.47
Creamed Coconut	3 lb	3.30
Sugar	10 lb	10.99
Corn syrup	13 lb	14.29
Ground whole fresh oranges (juice and rind)	50 lb	54.95
		Total 100.00

Procedure:

Place all ingredients in a clean kettle and cook to a medium jelly. Pour out on a tray evenly dusted with starch and allow to set. When set, cut into desired shapes and roll in a mixture of powdered sugar and starch.

Source: Kraft General Foods Corp.

Peco Flake

			% by Wt.
Granulated sugar	10	lb	32.10
Corn syrup	10	lb	32.10
Gem Slice or **Gem Chip Coconut**	6.5	lb	20.86
Salt	2.5	oz	0.50
Roasted no. 1 Spanish peanuts	4.5	lb	14.44
			Total 100.00

Procedure:

Place sugar and corn syrup in clean kettle together with enough water to dissolve sugar. Cook to 285 F.

Add coconut and salt. Continue cooking until coconut turns golden brown.

Add peanuts and mix well. Pour out on greased slab, spread to desired thickness. When cool enough, cut into bars or squares.

SUGGESTIONS:

Add dairy butter to improve quality. Add bicarbonate of soda if desired to make it lighter in texture.

Source: Kraft General Foods Corp.

Pineapple Coconut Creams

			% by Wt.
Gem Medium Coconut	2	lb	8.74
Invert sugar	8	oz	2.19
Water	4	oz	1.09
Fondant	10	lb	43.72
Creamed Coconut	1	lb	4.37
Sugar	6	lb	26.23
Corn syrup	1	lb	4.37
Frappé (q.v.)	2	lb	8.74
Convertit (invertase)	2	oz	0.55
Pineapple color and flavor			to suit
			Total 100.00

Procedure:

Place **Gem Medium Coconut**, melted invert sugar and water in mixer. Blend well.

Place fondant, **Creamed Coconut** and tenderized coconut in remelt kettle.

Combine sugar, corn syrup, and water to dissolve sugar in kettle. Cook to 244 F. Pour syrup over fondant and **Creamed Coconut**.

Lastly add frappé, **Convertit**, color, and flavor. Mix well. Cast in starch. When dry, shake out.

FRAPPÉ:		
Sugar	20	lb
Corn syrup	40	lb
Egg albumen (dissolved in water)	1	lb
Cold water	2	lb

Procedure:

Cook sugar to 232 F. Add corn syrup and place in beater. Add egg albumen which has been dissolved in water. Beat to a dry beat.

Source: Kraft General Foods Corp.

Pineapple Pecan Creams

			% by Wt.
Sugar	45	lb	62.40
Invert sugar	5	lb	6.94
Frappé (q.v.)	2	lb	2.77
Creamed Coconut	5	lb	6.94
Chopped pecans	5	lb	6.94
Crushed pineapple (drained)	5	lb	6.94
Convertit (invertase)	5	lb	6.94
Pineapple flavor and green color	1.5	oz	0.13
			Total 100.00

Procedure:

Add water to dissolve sugar. Bring to boil. Wash down sugar crystals from side of kettle. Cook to 244–246 F. Pour on open-type cream beater.

Cool batch to 100 F. Add color and start beater. When set, deposit or roll by hand. Dip in chocolate.

FRAPPÉ:

Sugar	20	lb
Corn syrup	40	lb
Egg albumen (dissolved in water)	1	lb
Cold water	2	lb

Procedure:

Add water to dissolve sugar. Cook to 232 F. Add corn syrup and place in beater. Add egg albumen which has been dissolved in water. Beat to a dry beat.

Source: Kraft General Foods Corp.

Plastic Coconut—Milk Centers

			% by Wt.
PART 1:			
Desiccated coconut (part medium, part macaroon)	20	lb	30.69
Water	2	lb	3.06
Condensed skimmed milk	5	lb	7.67
Nulomoline (invert sugar)	5	lb	7.67
Powdered salt	3	oz	0.29
PART 2:			
Granulated sugar	10	lb	15.34
Corn syrup	10	lb	15.34
Water	3	lb	4.60
PART 3:			
Basic Fondant (q.v.)	5	lb	7.67
Frappé No. 1 (q.v.)	5	lb	7.67
Orange, vanilla, raspberry, or lemon flavor			to suit
			Total 100.00

Procedure:

PART 1: Mix the water with the condensed milk, then add the coconut, **Nulomoline**, and salt and mix well. Set aside for 10 or 15 min.

PART 2: Boil sugar, corn syrup, and water to 250–255 F (sea level).

Pour the syrup (Part 2) into freshened coconut (Part 1), mixing thoroughly.

Add Part 3. Mix the batch well and, if desired, nuts, preserved fruit peels, raisins, etc. may be added.

Spread the batch out to cool, after which it may be formed by hand into pieces of various sizes, or it may be extruded from a rolled center machine in the form of large or small pieces.

The candy bars or smaller pieces may be partly coated with chocolate, placed into paper cups not coated, or completely coated with chocolate, thus producing small pieces suitable for package lines or larger pieces.

FRAPPÉ NO. 1:

Corn syrup	50	lb
Invert sugar	50	lb
Albumen	1	lb
Water	2	lb

Procedure:

Cook corn syrup to 245 F (sea level), shut off the heat and add invert sugar and stir until it is melted. Place the batch into a marshmallow beater, start the beater, then gradually add albumen which has been previously dissolved in water and beat the batch until light.

Source: Crompton & Knowles

Preserved Fruit–Coconut Slices

			% by Wt.
Corn syrup	20	lb	25.00
Sugar	12	lb	15.00
Nulomoline (invert sugar)	8	lb	10.00
Desiccated macaroon coconut	16	lb	20.00
Basic Fondant (q.v.)	8	lb	10.00
Frappé No. 1 (q.v.)	4	lb	5.00
Assorted preserved cherries, citron, pineapple, and diced lemon and orange peel	12	lb	15.00
Color and flavor			to suit
			Total 100.00

Procedure:

Place the corn syrup, sugar, **Nulomoline**, and water to dissolve the sugar into a kettle and bring to boiling point; then add coconut. Cook batch to a firm paste. Add fondnat, frappé, color and flavor as desired.

Spread half of batch out about $^3/_4$ in. high, cover batch on slab with preserved fruit and immediately cover preserved fruit with balance of batch. Roll down to the required height; when cold, cut the batch into slices, bars, or oblongs, and roll in fruit powdered sugar.

SUGGESTIONS:

Suggested flavor combinations: Rum and cinnamon, colored brown; vanilla and lemon, colored white.

For ripened banana flavor, add $2^1/_2$ lb of **Nulco MeloBan** to the batch after it has been cooked to a firm paste, plus banana color.

BASIC FONDANT:

Sugar	80 lb
Corn syrup	20 lb
Nulomoline	10 lb
Water	

Procedure:

Boil altogether to 244 F, cool and cream as usual.

FRAPPÉ NO. 1:

Corn syrup	50 lb
Invert sugar	50 lb
Albumen	1 lb
Water	2 lb

Procedure:

Cook corn syrup to 245 F (sea level), shut off the heat and add invert sugar and stir until it is melted. Place the batch into a marshmallow beater, start the beater, then gradually add albumen which has been previously dissolved in water and beat the batch until light.

Source: Crompton & Knowles

Rainbow Squares

			% by Wt.
Corn syrup	20	lb	16.49
Invert sugar	20	lb	16.49
Dextrose	5	lb	4.12
Water	1	lb	0.84
Salt	4	oz	0.21
Creamed Coconut	5	lb	4.12
Fondant 80-20 (q.v.)	20	lb	16.49
Gem Medium Coconut	40	lb	32.99
Frappé (q.v.)	10	lb	8.25
Color and flavor			to suit
			Total 100.00

Procedure:

Combine corn syrup, invert sugar, dextrose, water, and salt together in kettle and cook to 246–248 F.

Add **Creamed Coconut** and fondant. Stir until dissolved. Add **Gem Medium Coconut** and blend well. Lastly add frappé, color, and flavor. Mix thoroughly. Spread batch on oiled slab dusted with **Gem Macaroon Coconut**. Roll down to desired height. When cold, cut in pieces of desired size. Squares may be rolled in granulated sugar.

FONDANT 80-20:

Sugar	80	lb
Corn syrup	20	lb

Procedure:

Cook to 244 F.

FRAPPÉ:

Sugar	20	lb
Corn syrup	40	lb
Egg albumen (dissolved in water)	1	lb
Cold water	2	lb

Procedure:

Cook sugar to 232 F. Add corn syrup and place in beater. Add egg albumen which has been dissolved in water. Beat to a dry beat.

Source: Kraft General Foods Corp.

Slab Coconut Work

(For Cuts, Layers, Bars, Rolls, Etc.)

			% by Wt.
Corn syrup	15	lb	12.67
Nulomoline (invert sugar)	15	lb	12.67
Granulated sugar	20	lb	16.90
Water	3	lb	2.53
Medium shred or macaroon coconut	35	lb	29.57
Basic Fondant (q.v.)	20	lb	16.90
Frappé No. 1 (q.v.)	10	lb	8.45
Powdered salt	6	oz	0.31
Color, flavor, nuts, preserved fruits, grated lemon and orange peel, small gum drops, etc.			to suit
			Total 100.00

Procedure:

Place the corn syrup, **Nulomoline**, sugar, and water into a kettle and cook to 242–244 F. Add the coconut and mix well. Add the fondant, mix until the fondant is melted. Add the frappé and salt and mix thoroughly. If desired, add fruits, nuts, etc.

Spread the batch on an oiled and dusted slab. Roll to the required height and, when cold, cut into pieces.

THREE-LAYER CUTS AND BARS:

Divide the batch into three parts, coloring one portion pink, the second portion yellow, and the third portion to remain white.

Spread the batch into three sheets, placing them together. When cold, cut into bars, oblongs, or squares. Roll the pieces in XXX sugar and pack.

PECAN TOP COCONUT SLAB WORK:

Color a portion of the batch pink, adding Roman punch flavor. Spread the batch roughly on the slab about 4 in. high. Sprinkle pecan halves or pieces on top and then roll the batch down to about $^7/_8$ in. high. When cold, cut into squares, oblongs, or bars.

BASIC FONDANT:

Granulated sugar	80 lb
Corn syrup	20 lb
Invert sugar	10 lb
Water	

Procedure:

Boil granulated sugar, corn syrup, and invert sugar, plus sufficient water to dissolve the sugar, to 242 F (sea level), then cool and cream into fondant.

FRAPPÉ NO. 1:

Corn syrup	50 lb
Invert sugar	50 lb
Albumen	1 lb
Water	2 lb

Procedure:

Cook corn syrup to 245 F (sea level), shut off the heat and add invert sugar and stir until it is melted. Place the batch into a marshmallow beater, start the beater, then gradually add albumen which has been previously dissolved in water and beat the batch until light.

Source: Crompton & Knowles

Sliced Coconut Molasses Chews

		% by Wt.
Corn syrup	45 lb	22.64
Granulated sugar	15 lb	7.55
Molasses (**Grandma's**)	45 lb	22.64
Desiccated coconut (half macaroon, half medium or long shred)	60 lb	30.19
Gelatin (# 200 bloom, mixed with the water)	12 oz	0.38
Hot water	1.5 lb	0.75
Sweetened condensed whole milk	30 lb	15.10
Salt	1.5 lb	0.75
		Total 100.00

Procedure:

Place the corn syrup, sugar, molasses, condensed milk, and salt into a caramel cooking kettle, heat the batch until it boils, then add the desiccated coconut and continue to stir and cook the batch to a medium firm ball.

Add the gelatin-water combination, mix thoroughly, and spread the batch on an oiled slab lightly sprinkled with fine granulated sugar. Roll the batch down to form a sheet approximately 1 in. thick, sprinkling the surface with granulated sugar.

When the batch is firm, it may be cut into slices, oblongs, or squares and rolled in fine granulated sugar.

Brazil nuts, peanuts, almonds, or walnuts may be added to the above batch if desired.

Source: Crompton & Knowles

Southern Flips

			% by Wt.
PART 1:			
Coconut (part macaroon, part medium shred)	35	lb	29.17
Molasses (**Grandma's**)	20	lb	16.67
PART 2:			
Corn syrup	10	lb	8.33
Nulomoline (invert sugar)	10	lb	8.33
Granulated sugar	25	lb	20.83
Water (to dissolve sugar)			
Dairy butter	5	lb	4.17
Basic Fondant (q.v.)	10	lb	8.33
Frappé No. 1 (q.v.)	5	lb	4.17
Powdered salt			to suit
			Total 100.00

Procedure:

PART 1: Mix together and set aside for at least 15 min.

PART 2: Boil the corn syrup, **Nulomoline**, sugar, water, and butter to 260–265 F (sea level), and pour on top of the molasses–coconut mixture (Part 1); then add the Basic Fondant, frappé, and salt, and mix well.

When the batch is partially cool, spread it out and cut into bars or pass through a rolled cream center machine, then coat with chocolate or icing.

BASIC FONDANT:

Granulated sugar	80 lb
Corn syrup	20 lb
Invert sugar	10 lb
Water	

Procedure:

Boil granulated sugar, corn syrup, and invert sugar, plus sufficient water to dissolve the sugar, to 242 F (sea level), then cool and cream into fondant.

FRAPPÉ NO. 1:

Corn syrup	50 lb
Invert sugar	50 lb
Albumen	1 lb
Water	2 lb

Procedure:

Cook corn syrup to 245 F (sea level), shut off the heat and add invert sugar and stir until it is melted. Place the batch into a marshmallow beater, start the beater, then gradually add albumen which has been previously dissolved in water and beat the batch until light.

Source: Crompton & Knowles

Special Coconut Bar

No. 1

		% by Wt.
PART 1:		
Desiccated coconut	65 lb	28.61
Nulomoline (invert sugar)	20 lb	8.80
Corn syrup	20 lb	8.80
Water	10 lb	4.40

PART 2:			
Refined sugar	50	lb	22.00
Corn syrup	30	lb	13.20
Water	12	lb	5.28
PART 3:			
Fondant 80-20	10	lb	4.40
Salt	4	oz	0.11
Standard Frappé (q.v.)	10	lb	4.40
			Total 100.00

Procedure:

PART 1: Heat the corn syrup, **Nulomoline**, and water to approximately 150 F. Add the coconut and mix thoroughly. Permit the batch to remain undisturbed for at least 1 h so as to give the coconut time to absorb the moisture.

PART 2: Boil altogether to 250 F.

Pour Part 2 at once into Part 1, mixing thoroughly. Then add the Fondant 80-20 which has been cooked to 244 F. Also add salt and Standard Frappé. Mix well.

When partially cooled, form into bars, cakes, kisses, etc. of desired size.

STANDARD FRAPPÉ:		
Corn syrup	50	lb
Nulomoline (invert sugar)	50	lb
Albumen (soaked in water)	1	lb
Cold water	2	lb

Procedure:

Boil the corn syrup to 245 F. Add the **Nulomoline** and mix until melted. Place this at once into a marshmallow beater, start beater in motion, gradually add dissolved albumen and beat until quite light. Set aside for future use.

Source: Crompton & Knowles

No. 2

(Desiccated Coconut)

			% by Wt.
PART 1:			
Desiccated coconut	65	lb	28.58
Nulomoline (invert sugar)	20	lb	8.80
Corn syrup	20	lb	8.80
Water	10	lb	4.40
PART 2:			
Refined sugar	50	lb	21.99
Corn syrup	30	lb	13.19
Water	12	lb	5.28
PART 3:			
Fondant 80-20	10	lb	4.40
Salt	4	oz	0.11
Convertit (invertase)	2	oz	0.05
Frappé 50-50 (q.v.)	10	lb	4.40
			Total 100.00

Procedure:

PART 1: Heat the corn syrup, **Nulomoline**, and water to approximately 150 F. Add the coconut and mix thoroughly. Permit this batch to lay undisturbed for at least $^1/_2$ h so as to give the coconut time to absorb the moisture.

PART 2: Boil altogether to 250 F.

Pour Part 2 at once into Part 1, mixing thoroughly, then add Fondant 80-20 which has been cooked to 244 F. Now add salt, **Convertit**, and Frappé 50-50.

FRAPPÉ 50-50:		
Nulomoline (invert sugar)	50	lb
Corn syrup	50	lb
Albumen (dissolved in water)	1	lb
Cold water	2	lb

Procedure:

Cook corn syrup to a temperature of 245 F. Shut off the steam and add cold **Nulomoline** and stir until the latter is melted.

Place the batch into a marshmallow beater, start the machine and add the albumen which has been previously dissolved. Beat until light and use as needed.

Source: Crompton & Knowles

Special Rolled Coconut Creams

(For Bars)

			% by Wt.
Coconut paste (q.v.)	8	lb	10.25
Fondant (q.v.)	65	lb	83.26
Convertit	1	oz	0.08
Frappé	5	lb	6.41
Vanilla			to suit
			Total 100.00

Procedure:

Prepare coconut paste and fondant. Spread about 8 lb of the coconut paste over the ball beater. Pour fondant batch on top of coconut paste and sprinkle batch with not more than 2 oz of water. Allow batch to cool to approximately 100 F.

Now start machine in motion, add 1 oz **Convertit**, and permit batch to beat until a slight grain becomes visible; then add 5 lb of frappé, sufficient vanilla, scraping sides and mixing arms of beater; and when batch turns up firm, stop the beater. Place batch on cooling slabs so that it will cool to room temperature. Do not attempt to run the batch through the machine until it is cool.

COCONUT PASTE:

Long shred coconut	12	lb
Nulomoline (invert sugar)	18	lb
Water	2	lb

Procedure:

Place the water and **Nulomoline** into a kettle and heat until the **Nulomoline** is completely dissolved (do not permit the batch to boil). Add coconut and mix thoroughly. Heat batch slightly for a minute or two and set aside for at least 2 h.

It is well to keep a stock of this preparation on hand as it is the function of **Nulomoline** to soften coconut, thus making it tender and preventing it from absorbing moisture from the fondant.

FONDANT:

Refined cane sugar	44	lb
Nulomoline (invert sugar)	6	lb
Water	15	lb

Procedure:

Place water, sugar, and **Nulomoline** into a clean kettle. Stir batch occasionally until it reaches boiling point. Wash down sides of kettle to remove all grains of sugar. Place wooden cover on top of kettle and permit batch to boil rapidly for a few minutes; then partly remove cover and insert a tested thermometer. Boil batch as rapidly as possible to 250 F. Turn off the heat.

Source: Crompton & Knowles

Tang of the Tropics

(Molasses Coconut Squares, Oblongs, Slices, and Clusters)

			% by Wt.
Corn syrup	20	lb	22.13
Molasses (**Grandma's**)	20	lb	22.13
Medium shred or macaroon coconut	30	lb	33.20
Basic Fondant (q.v.)	20	lb	22.13
Powdered salt	6	oz	0.41
			Total 100.00

Procedure:

Place the corn syrup and molasses into a kettle, bring the batch to boiling point. Add the coconut. Mix constantly and cook the batch to a medium firm ball. Add the Basic Fondant, salt, and mix thoroughly, then handle the batch in the following manner:

Turn batch out on an oiled slab and sprinkle with XXX sugar. Roll batch down to various heights; cut into squares, oblongs, slices, etc. which should be rolled in XXX sugar.

BASIC FONDANT:

Granulated sugar	80 lb
Corn syrup	20 lb
Nulomoline (invert sugar)	10 lb
Water (to dissolve the sugar)	

Procedure:

Heat altogether, stirring occasionally until the batch boils, then wash down all grains of sugar that adhere to inside of the kettle; boil batch rapidly to 244–246 F. Cool syrup to 110–120 F and beat into fondant.

BRAZIL NUTS BARS AND CUTS:

Add whole or broken Brazils to a portion of the "Tang of the Tropics" batch, roll down to about $^1/_2$ in. in height and when cold, cut into bars about $4^1/_2$ in. long and $^1/_2$ in. thick.

A portion of the batch may also be cut into oblongs or squares, and the pieces rolled in XXX sugar.

TOASTED PEANUT COCONUT BARS AND CUTS:

Take a portion of the "Tang of the Tropics" batch, add toasted peanuts, mix well and spread batch on a slab well sprinkled with coarse toasted coconut. Roll the batch down to about $1^1/_2$ in. high, then sprinkle the surface of the batch with toasted coconut and roll down to the desired height. Cut into bars or smaller pieces.

TOASTED COCONUT MOLASSES ROLLS:

Cook a batch of "Tang of the Tropics" to a medium soft paste and form it into rolls which should be later moistened with syrup, then coated in chocolate icing and rolled in toasted coconut.

MINT MOLASSES CUTS:

Spread a portion of the "Tang of the Tropics" batch on an oiled slab dusted with XXX sugar. Roll it down to about $^3/_4$ in. thick, then spread a layer of "Variety Slab Coconut No. 1" batch, which has been colored green and flavored with mint oil, on top of the molasses coconut sheet and roll down to the required height. When cold, cut into slices, squares, oblongs, or bars.

Source: Crompton & Knowles

Toasted Coconut Peanut Niblets

			% by Wt.
Granulated sugar	30	lb	37.50
Corn syrup	30	lb	37.50
Peanut butter	20	lb	25.00
		Total	100.00

Procedure:

Place sugar and corn syrup in a kettle with water to dissolve sugar. Cook to 290–300 F. Set aside a small amount of clear candy for jacket. Pull remainder of batch up light and spread out and place 20 lb of warm peanut butter in center of jacket and honeycomb eight times. Now cover with clear candy jacket. Run on peachstone rolls and coat with **Golden Toasted Macaroon Coconut** in a revolving pan. Use syrup to make coconut stick. When dry, pack.

SYRUP FOR COATING PEACHSTONE ROLLS:

Sugar	2	lb
Corn syrup	1	lb

Procedure:

Cook to 220 F.

Source: Kraft General Foods Corp.

Variety Slab Coconut Work

No. 1

(For Cuts, Layers, Bars, Rolls, Croquettes, Cakes, Etc.)

			% by Wt.
Corn syrup	20	lb	17.96
Nulomoline (invert sugar)	20	lb	17.96
Granulated sugar	5	lb	4.49
Water	1	lb	0.90
Medium shred or macaroon coconut	35	lb	31.41
Basic Fondant (q.v.)	20	lb	17.96
Frappé No. 1	10	lb	8.98

Powdered salt	6 oz	0.34
Color, flavor, nuts, preserved fruits, ground lemon and orange peel, small gum drops, etc.		to suit
		Total 100.00

Procedure:

Place the corn syrup, **Nulomoline**, sugar, and water into a kettle and boil altogether to 244–246 F. Add the desiccated coconut, mix well, add the Basic Fondant and mix until the fondant is dissolved, then add the frappé and salt, mix thoroughly, and, if desired, add fruits, nuts, etc.

Spread the batch on an oiled and dusted slab, roll down to the required height and when cold, cut into pieces of the desired size.

BASIC FONDANT:

Granulated sugar	80 lb
Corn syrup	20 lb
Nulomoline (invert sugar)	10 lb
Water (to dissolve the sugar)	

Procedure:

Heat altogether, stirring the batch occasionally until it boils, then wash down all grains of sugar that adhere to the inside of the kettle and boil the batch rapidly to 244–246 F. Cool the syrup to 110–125 F, then beat into fondant.

FRAPPÉ NO. 1:

Corn syrup	50 lb
Invert sugar	50 lb
Albumen	1 lb
Water	2 lb

Procedure:

Cook corn syrup to 245 F (sea level), shut off the heat and add invert sugar and stir until it is melted. Place the batch into a marshmallow beater, start the beater, then gradually add albumen which has been previously dissolved in water and beat the batch until light.

Source: Crompton & Knowles

No. 2

			% by Wt.
Corn syrup	20	lb	15.83
Invert sugar	20	lb	15.83
Sugar	5	lb	3.96
Water	1	lb	0.79
Creamed Coconut	15	lb	11.87
Basic Fondant (q.v.)	20	lb	15.83
Salt (powdered)	6	oz	0.30
Gem Medium or **Gem Macaroon Coconut**	35	lb	27.68
Nougat cream	10	lb	7.91
			Total 100.00

Procedure:

Combine corn syrup, invert sugar, sugar, and water in steam kettle and cook to 244–246 F.

Add **Creamed Coconut**, salt, and Basic Fondant. Stir until dissolved.

Next add **Gem Medium Coconut** or **Macaroon** and mix well. Lastly add nougat cream and mix thoroughly. Spread batch on an oiled, dusted slab. Roll to required height. When cold, cut to desired size.

BASIC FONDANT:

Sugar	80	lb
Corn syrup	20	lb
Nulomoline	10	lb
Water (to dissolve sugar)		

Procedure:

Cook to 244–246 F.

Source: Kraft General Foods Corp.

Walnut Coconut Light Divinity

			% by Wt.
Invert sugar	10	lb	12.10
Vegetable albumen (soy albumen)	1	lb	1.20
Sugar (6X)	2	lb	2.42

Water	2	lb	2.42
Sugar	35	lb	42.32
Corn syrup	15	lb	18.15
Fondant 80-20 (q.v.)	10	lb	12.10
Salt	1	oz	0.08
Convertit (invertase)	1	oz	0.08
Creamed Coconut	8	oz	0.61
Chopped walnuts	2	lb	2.42
Vanilla flavor	5	lb	6.10
			Total 100.00

Procedure:

Combine 6X sugar and soy albumen. Blend well. Add water and mix for 5 min. Heat invert sugar to 150 F. Combine invert sugar and albumen solution. Place in mixer and beat until light.

Place in kettle with water the second amount of sugar to dissolve. Cook to 245–250 F. Now pour this into the whipped portion, beating continuously.

Add fondant, salt, **Convertit**, **Creamed Coconut**, chopped walnuts, and vanilla flavor and mix thoroughly into batch.

FONDANT 80-20:

Sugar	80	lb
Corn syrup	20	lb

Procedure:

Cook to 244 F.

Source: Kraft General Foods Corp.

Chapter 7

DIETETIC CANDIES

People prone to obesity and those suffering from chronic illnesses such as diabetes or a coronary condition must limit their intake of candies and confections. The ingredients under taboo are sugars, dextrose, and starch; and it is these ingredients that dietetic confections seek to replace. The use of sorbitol has increased as a sugar substitute, especially since other artificial sweeteners have come under question. Dietetic candies and confections include diabetic chocolates, slenderizing chocolates, and sugar-free gums.

Sugar-free Hard Gums

			% by Wt.
PART 1:			
Genu Pectin Type X-5431	4.500	kg	3.15
Sodium carbonate	0.075	kg	0.05
Water	45.000	kg	31.47
PART 2:			
Sorbitol syrup	93.400	kg	65.33
PART 3:			
Color and flavor			to suit
		Total	100.00
		Evaporation	30.00
		Yield	70.00

Source: Copenhagen Pectin Factory Ltd.

Sugar-free Fruit Jellies

			% by Wt.
PART 1:			
Water	25.000	l	18.79
Sodium citrate • $2H_2O$	0.400	kg	0.30
Citric acid (50% w/v)	0.500	l	0.38
PART 2:			
Genu Pectin Type X-5431	1.200	kg	0.90
Sorbitol	5.000	kg	3.75
PART 3:			
Sorbitol syrup (70% SS)	100.000	kg	75.16
Artificial sweetener			to suit
PART 4:			
Citric acid (50% w/v)	0.900	l	0.68
Sodium citrate • $2H_2O$	0.050	kg	0.04
			Total 100.00
			Evaporation 25.00
			Yield 75.00

pH before addition of acid (Part 4) — 4.3
pH in the product — 3.7

NOTES:

Sodium citrate (Part 4) must be dissolved in the citric acid (Part 4) before this is added to the batch, otherwise local pregelation is liable to occur.

Source: The Copenhagen Pectin Factory Ltd.

Chapter 8

FUDGE

Fudge for Casting into Starch

(Small Shapes)

			% by Wt.
Corn syrup	35	lb	22.99
Nulomoline (invert sugar)	5	lb	3.28
Condensed whole milk (containing 40% sugar)	20	lb	13.14
Refined sugar	50	lb	32.84
Coconut butter (86 F m.p.)	2	lb	1.32
Fondant (q.v.)	35	lb	22.99
Fudge Frappé (q.v.)	5	lb	3.28
Salt	4	oz	0.16
Flavor			to suit
			Total 100.00

Procedure:

Place into a kettle equipped with a good stirring device the corn syrup, **Nulomoline**, condensed milk, and sugar, then add 8 lb of water (1 gal). Turn the steam on partly and when the batch reaches the boiling point wash down all grains of sugar that adhere to the sides of the kettle and mixing device; then boil the batch rapidly to 238 F.

Turn the steam off, allow the batch to mix for 3 min, then add the fondant and coconut butter and mix well. Then add the frappé, salt, and flavor. Mix thoroughly and place the fudge into the hopper of the depositor. The hopper should be heated to approximately 175 F.

Cast the fudge into small impressions and after the fudge has set it may be removed from the starch, thoroughly cleaned then placed into a revolving pan and coated with lac. Spread the fudge pieces out to dry after which they are ready to pack into boxes or pails.

FONDANT:

Sugar	80 lb
Water	20 lb
Corn syrup	20 lb

Procedure:

Cook to 244 F.

FUDGE FRAPPÉ:

Corn syrup	50 lb
Nulomoline	50 lb
Albumen (dissolved in water)	1.5 lb
Water	3 lb

Procedure:

Boil corn syrup to 250 F. Add **Nulomoline** and mix well. Turn off the heat and place this into a marshmallow beater, then gradually add albumen dissolved in water and beat until light.

Source: Crompton & Knowles

Chocolate Fudge

			% by Wt.
Coffee sugar	6	lb	63.58
Unsweetened chocolate	2	lb	21.19
Sweet cream	2	c	11.26
Warmed butter	4–5	oz	3.31
Vanilla	3–4	Tbl	0.66
		Total	100.00

Procedure:

To coffee sugar, add best quality unsweetened chocolate, grated. Stir until thoroughly mixed then add of sweet cream, beat to a smooth paste, adding as you proceed, the warmed butter. This done, put the mixture in a kettle where it will melt very slowly (it should not reach the boil under 2 h).

At the end of that time, increase the heat and boil fast for 10 min or until it reaches "the ball;" then stir in vanilla sufficient to flavor well, and pour on slab, or into greased pans, and while still soft, mark into squares.

Source: Savage Brothers Co.

Light Chocolate Fudge

			% by Wt.
Granulated sugar	36	lb	24.43
Corn syrup	24	lb	16.28
Nulomoline (invert sugar)	3	lb	2.04
Coconut oil (76 F m.p.)	3	lb	2.04
Sweetened condensed whole milk	27	lb	18.32
Water	3	lb	2.04
Fondant (q.v.)	24	lb	16.28
Chocolate liquor	6	lb	4.07
Cocoa powder (18–20% butterfat)	6	lb	4.07
Frappé No. 2 (q.v.)	15	lb	10.18
Salt	6	oz	0.25
Vanilla flavor			to suit
			Total 100.00

Procedure:

Place the sugar, corn syrup, coconut oil, **Nulomoline**, condensed milk, and water into a kettle. Mix, apply heat and with constant stirring cook to 242 F. Remove from heat, add the liquor chocolate, sift in the cocoa, add the fondant, and mix well. Add the frappé, salt, and flavor. Mix and stir until the batch becomes heavy, pour on oiled waxed paper.

FONDANT:

Granulated sugar	100	lb
Corn syrup	20	lb
Nulomoline (invert sugar)	10	lb

Procedure:

Cook to 245 F, cool and beat into fondant.

FRAPPÉ NO. 2:

Nulomoline (invert sugar)	50	lb
Corn syrup	50	lb
Egg albumen (spray-dried powder)	2	lb

Procedure:

Place half the **Nulomoline** and egg albumen into a marshmallow beater. Mix at slow speed. Meanwhile, heat the remaining **Nulomoline** and corn syrup to approximately 200 F. Add to the mixture in the beater. Beat until light.

Source: Crompton & Knowles

Vanilla Plastic Fudge

			% by Wt.
Granulated sugar	30	lb	24.39
Water	3	lb	2.44
Nulomoline (invert sugar)	5	lb	4.07
Corn syrup	15	lb	12.20
Condensed skimmed milk	30	lb	24.39
Hard coconut butter (110 F m.p.)	5	lb	4.07
Frappé No. 1 (q.v.)	5	lb	4.07
Basic Fondant (q.v.)	20	lb	16.26
Cashew nuts	5	lb	4.05
Pecans	5	lb	4.06
Salt (powdered)			to suit
Vanilla			to suit
			Total 100.00

Procedure:

Place the sugar, water, **Nulomoline**, corn syrp, and condensed milk into a kettle and boil altogether to approximately 240 F (sea level).

Drop into the batch the coconut butter and Basic Fondant but do not mix until the batch cools to approximately 175 F, then add the salt, vanilla, frappé, and nuts and mix thoroughly. Spread the batch out to set.

FRAPPÉ NO. 1:

Corn syrup	50 lb
Nulomoline (invert sugar)	50 lb
Albumen (dissolved in water)	1 lb
Water	2 lb

Procedure:

Boil the corn syrup to 245 F, shut off the heat and add the **Nulomoline** and stir until it is melted. Place the batch into a marshmallow beater, start the beater, then gradually add the albumen which has been previously dissolved in the water and beat the batch until light.

BASIC FONDANT:

Granulated sugar	80 lb
Corn syrup	20 lb
Invert sugar	10 lb
Water	

Procedure:

Boil granulated sugar, corn syrup, and invert sugar, plus sufficient water to dissolve the sugar, to 242 F (sea level), then cool and cream into fondant.

Source: Crompton & Knowles

Chocolate Plastic Fudge

		% by Wt.
Granulated sugar	30 lb	24.00
Water	5 lb	4.00
Nulomoline (invert sugar)	7 lb	5.60
Corn syrup	15 lb	12.00
Condensed skimmed milk	20 lb	16.00
Cocoa powder	9 lb	7.20
Frappé No. 1 (q.v.)	4 lb	3.20
Basic Fondant	25 lb	20.00
Cashew nuts	5 lb	4.00
Pecans	5 lb	4.00

Salt (powdered)	to suit
Vanilla	to suit
	Total 100.00

Procedure:

Place the sugar, water, **Nulomoline**, corn syrup, condensed milk, and cocoa powder into a kettle and boil altogether to approximately 237 F (sea level).

Drop into the batch the Basic Fondant but do not mix until the batch cools to approximately 175 F, then add the salt, vanilla, frappé, and nuts and mix thoroughly. Spread the batch out to set.

FRAPPÉ NO. 1:

Corn syrup	50 lb
Nulomoline (invert sugar)	50 lb
Albumen (dissolved in water)	1 lb
Water	2 lb

Procedure:

Boil the corn syrup to 245 F, shut off the heat and add the **Nulomoline** and stir until it is melted. Place the batch into a marshmallow beater, start the beater, then gradually add the albumen which has been previously dissolved in the water and beat the batch until light.

BASIC FONDANT:

Granulated sugar	80 lb
Corn syrup	20 lb
Invert sugar	10 lb
Water	

Procedure:

Boil granulated sugar, corn syrup, and invert sugar, plus sufficient water to dissolve the sugar, to 242 F (sea level), then cool and cream into fondant.

Source: Crompton & Knowles

Special Chocolate Fudge

			% by Wt.
Granulated sugar	12	lb	25.00
Corn syrup	8	lb	16.67
Butter	2	lb	4.17
Water	2	lb	4.17
Liquor chocolate	4	lb	8.33
Fondant	20	lb	41.66
			Total 100.00

Procedure:

Cook the sugar, corn syrup, butter, and water to 244 F, then add the liquor chocolate and the fondant, mixing well and permit the batch to cool to approximately 175 F. Add salt, vanilla, nut meats, etc., and spread the batch into trays covered with oiled waxed paper.

Source: Crompton & Knowles

Chocolate Almond Fudge

			% by Wt.
Granulated sugar	40	lb	18.43
Cocoa powder	15	lb	6.90
Water	1	gal	3.85
Corn syrup	30	lb	13.82
Invert sugar	10	lb	4.60
Sweetened condensed milk	40	lb	18.43
Hard coconut butter	8	lb	3.69
Salt	12	oz	0.35
Casting fondant	40	lb	18.43
Frappé whip	10	lb	4.60
Mission almonds (select sheller run, 27/30 count, roasted)	15	lb	6.90
Vanilla flavor			to suit
			Total 100.00

Procedure:

Mix granulated sugar and cocoa powder in double-action cooking kettle, add water, corn syrup, invert sugar, and condensed milk. Cook to 242–246 F. Cool to 200 F, add coconut butter, salt, and fondant, mixing thoroughly until fondant is melted. Mix in frappé, flavor, and almonds. Spread on waxed paper to thickness of about 2 in. Almonds may also be imbedded in the surface. When set, cut into strips or bars exposing the almonds to view.

Source: California Almond Growers Exchange

Chocolate Slab Fudge

No. 1

			% by Wt.
Granulated sugar	60	lb	25.46
Corn syrup	40	lb	16.98
Nulomoline (invert sugar)	5	lb	2.12
Coconut oil (76 F m.p.)	5	lb	2.12
Condensed milk	45	lb	19.10
Basic Fondant (q.v.)	40	lb	16.98
Cocoa powder	15	lb	6.37
Powdered salt	10	oz	0.27
Frappé No. 2 (q.v.)	25	lb	10.60
Plum flavor			to suit
			Total 100.00

Procedure:

Place the sugar, corn syrup, **Nulomoline**, coconut oil, condensed milk, and 6 lb of water into a kettle and boil altogether to 242 F (sea level). Turn off the heat, add the Basic Fondant and cocoa powder and mix well, then add the salt, frappé, and flavor and mix thoroughly. As soon as the batch becomes rather heavy, spread it on waxed paper. When the surface of the batch has partially set, work into it ground walnut meats.

BASIC FONDANT:

Granulated sugar	80	lb
Corn syrup	20	lb
Invert sugar	10	lb
Water		

Procedure:

Boil granulated sugar, corn syrup, and invert sugar, plus sufficient water to dissolve the sugar, to 242 F (sea level), then cool and cream into fondant.

FRAPPÉ NO. 2:

Corn syrup	50	lb
Nulomoline (invert sugar)	50	lb
Albumen (dissolved in water)	2	lb
Cold water	4	lb

Procedure:

Boil the corn syrup to approximately 250 F (sea level). Turn off the heat, add the **Nulomoline** and stir until it is melted. Place the batch into a marshmallow beater, start the beater then gradually add the albumen solution and beat the batch until light.

Source: Crompton & Knowles

No. 2

(Dri-Fond)

			% by Wt.
Granulated sugar	40	lb	29.68
Corn syrup	20	lb	14.84
Nulomoline (invert sugar)	5	lb	3.71
Unsweetened evaporated milk	25	lb	18.55
Cocoa butter	5	lb	3.71
Unsweetened liquor chocolate	5	lb	3.71
Dri-Fond (dry fondant)	20	lb	14.84
Water	2	lb	1.48
Frappé No. 2 (q.v.)	12.5	lb	9.28
Salt	4	oz	0.20
Vanilla or other flavor			to suit
			Total 100.00

Procedure:

Mix the **Dri-Fond** with water and set aside. Place the sugar, corn syrup,

Nulomoline, cocoa butter, liquor chocolate and half of the evaporated milk in a kettle. Apply heat and bring to the boiling point, stirring continuously. Gradually add the remaining evaporated milk with constant stirring and cook to 240 F.

Remove from heat, add the **Dri-Fond**/water mixture and mix well. Add the salt, frappé, and flavor and mix thoroughly.

As soon as the batch has become firm, spread on oiled waxed paper. When cool, the batch may be scored and cut into pieces of the desired size and shape.

FRAPPÉ NO. 2:

Corn syrup	25 lb
Nulomoline (invert sugar)	25 lb
Albumen (dissolved in the water)	1 lb
Water (cold)	2 lb

Procedure:

Boil the corn syrup to 245 F, add the **Nulomoline** and mix until it is completely dissolved. Turn off the heat, place the batch into a beating machine, start the beater, then gradually add the albumen solution, and beat the batch until it is quite stiff. Remove the frappé from the beater and place into a suitable container for future use.

Source: Crompton & Knowles

Commercial Fudge

	% by Wt.
Granulated sucrose	33.3
Clinton regular corn syrup (43 Bé)	16.6
Sweetened condensed skim milk	13.3
Hardened vegetable oil	1.9
Water	8.0
Fondant	20.0
Frappé	6.7
Salt	0.2
Vanilla flavor	to suit
	Total 100.0

Procedure:

Place sweetened condensed skim milk, sugar, corn syrup, vegetable oil, and water in steam-jacketed kettle equipped with agitator.

Cook to 242–244 F. Draw off batch into another kettle. Add fondant, frappé, salt, and flavor. If making chocolate fudge, add chocolate liquor at this point (see note). Mix until graining starts.

Pour on table or trays lined with oiled paper. Allow to cool and score pieces. When firm enough to handle, turn over, remove paper, and allow exposed surface to dry. Pack.

NOTE:

A chocolate fudge can be made with this formula by adding approximately 7 lb of chocolate liquor.

Source: ADM

Cast Fluffy Fudge

			% by Wt.
Granulated sugar	30	lb	19.74
Sweetened condensed whole milk	25	lb	16.45
Corn syrup	20	lb	13.16
Coconut oil (76 F m.p.)	5	lb	3.29
Water	2	lb	1.32
Basic Fondant (q.v.)	50	lb	32.88
Frappé No. $1^1/_2$ (q.v.)	20	lb	13.16
Salt and butter-vanilla flavor			to suit
			Total 100.00

Procedure:

Place the sugar, condensed milk, corn syrup, coconut oil, and water in a caramel cooking kettle. Apply heat and cook to 242 F. Add the fondant, mix well. Add the frappé, salt, and flavor. Mix well and cast into starch. When set, remove from starch and coat with chocolate or hand icing.

BASIC FONDANT:

Granulated sugar	80	lb
Corn syrup	20	lb
Invert sugar	10	lb
Water		

Procedure:

Boil granulated sugar, corn syrup, and invert sugar, plus sufficient water to dissolve the sugar, to 242 F (sea level), then cool and cream into fondant.

FRAPPÉ NO. $1^1/_2$:

Nulomoline (invert sugar)	50 lb
Corn syrup	50 lb
Egg albumen (spray-dried powder)	1.5 lb

Procedure:

Place half the **Nulomoline** and egg albumen into a marshmallow beater. Mix at slow speed. Meanwhile, heat the remaining **Nulomoline** and corn syrup to approximately 200 F. Add to the mixture in the beater. Beat until light.

Source: Crompton & Knowles

Fruit–Fudge–Pudding

(Cut or Rolled)

			% by Wt.
Sugar	30	lb	21.90
Water	2	lb	1.46
Condensed skimmed or whole milk	20	lb	14.60
Corn syrup	20	lb	14.60
Nulomoline (invert sugar)	5	lb	3.65
Hard coconut butter	3	lb	2.19
Basic Fondant (q.v.)	25	lb	18.25
Frappé No. 2 (q.v.)	20	lb	14.60
Salt			to suit
Rum flavor			to suit
Assorted preserved fruits (glaced cherries, pineapple, orange peel, etc.)	12	lb	8.75
			Total 100.00

Procedure:

Place the corn syrup into a kettle equipped with mixers. Add the coconut butter, **Nulomoline**, condensed milk, sugar, and water, turn on the heat and cook the batch, stirring constantly, to 245 F if the batch is to be rolled by hand, or cook to 250 F if it is to be spread out and cut like fudge.

As soon as the batch has been cooked to the proper consistency, turn off the heat, add the salt and Basic Fondant, mixing thoroughly, then add the flavor and frappé, and mix well. Later add the preserved fruits which should be properly blended with the batch.

When the batch is cold, roll or cut into bars of the desired size and coat with chocolate.

NOTE:

A portion of the preserved fruits may be pressed into the surface of the batch after it has been spread out on waxed paper to set.

BASIC FONDANT:

Granulated sugar	80 lb
Corn syrup	20 lb
Invert sugar	10 lb
Water	

Procedure:

Boil granulated sugar, corn syrup, and invert sugar, plus sufficient water to dissolve the sugar, to 242 F (sea level), then cool and cream into fondant.

FRAPPÉ NO. 2:

Corn syrup	25 lb
Nulomoline (invert sugar)	25 lb
Albumen (dissolved in the water)	1 lb
Water (cold)	2 lb

Procedure:

Boil the corn syrup to 245 F, add the **Nulomoline** and mix until it is completely dissolved. Turn off the heat, place the batch into a beating machine, start the beater, then gradually add the albumen solution, and beat the batch until it is quite stiff. Remove the frappé from the beater and place into a suitable container for future use.

Source: Crompton & Knowles

Good Grade Fudge

(For "Pan," Bars, or Small Pieces)

		% by Wt.
Sugar	7 lb	38.89
Corn syrup	2 lb	11.11
Nulomoline (invert sugar)	1 lb	5.56
Cream (20% butterfat)	2 lb	11.11
Evaporated milk	1 lb	5.56
Fondant (q.v.)	2.5 lb	13.89
Frappé No. 1 (q.v.)	1 lb	5.56
Chocolate (unsweetened)	1.5 lb	8.32
Nut meats		to suit
		Total 100.00

Procedure:

Cook sugar, corn syrup, **Nulomoline**, cream, and evaporated milk altogether to 238–240 F. Add the fondant, and the remainder of the ingredients.

NOTE:

To make vanilla fudge, leave out the chocolate and cook the batch to 240 F.

FONDANT:

Sugar	40 lb
Corn syrup	5 lb
Nulomoline (invert sugar)	5 lb
Water (to dissolve the sugar)	

Procedure:

Cook to 240–242 F. Cool and beat into fondant.

FRAPPÉ NO. 1:

Corn syrup	25 lb
Nulomoline (cold)	25 lb
Albumen (dissolved in water)	8 oz
Water	1 lb

Procedure:

Cook the corn syrup to 245 F. Shut off the steam and add the cold **Nulomoline** and stir until it is melted. Place the batch into a marshmallow beater, start the beater, then gradually add the albumen which has previously been dissolved in water. Beat the batch until light and use as directed.

Source: Crompton & Knowles

Maple Fudge

No. 1

			% by Wt.
Maple sugar syrup	90	lb	85.71
Maple sugar invert syrup	10	lb	9.53
Maple cream	5	lb	4.76
		Total	100.00

Procedure:

Boil together to 240 F, adding any desired amount of cream or evaporated milk. Continue to boil the batch to 242 F, then cool it to approximately 170 F. Add maple cream, mix well, and cast into starch or rubber molds or spread it out on waxed paper to set.

Source: Crompton & Knowles

No. 2

			% by Wt.
Maple sugar	1	lb	47.06
Milk	1	pt	50.00
Baking soda		pinch	0.14
Butter	1	Tbl	2.80
		Total	100.00

Procedure:

Crush a quantity of maple sugar fine with rolling pin, and use in proportion of 1 lb sugar to 1 pt of milk. Pour milk hot over the sugar, having previously added a pinch of baking soda to the milk. Let the sugar melt slowly over the fire, and then bring to a fast boil. When the syrup threads from the tip of a spoon, stir in a tablespoonful of butter for each pound of sugar. When it again reaches the boil, pour out on slab, or into buttered pans, and cut into squares as it hardens.

Source: Savage Brothers Co.

Penocha Fudge

			% by Wt.
Brownulated powdered brown sugar	25	lb	18.50
Invert sugar	25	lb	18.50
Cream (30% butterfat)	10	lb	7.40
Regular corn syrup	10	lb	7.40
Salt	2.5	oz	0.11
Amerfond fondant sugar	50	lb	36.99
Melted vegetable fat (92 F m.p.)	5	lb	3.70
Chopped pecans	10	lb	7.40
			Total 100.00

Procedure:

Place powdered brown sugar, invert sugar, cream, corn syrup, and salt in cooking kettle and bring to boil. Transfer to remelt kettle or double boiler. Add fondant sugar and stir until smooth. Add melted vegetable fat and mix well. Add chopped pecans and mix well.

Pour at 160 F onto paper-lined tables or pans.

Source: Amstar

Plain Fudge

			% by Wt. (with cream)
Sugar	12	lb	32.88
Glucose (corn syrup)	8	lb	21.92
Water	2	qt	10.96

or { Cream	4	qt	23.28
Condensed milk }	2	qt	—
Cream fondant	4	lb	10.96
			Total 100.00

Procedure:

Place sugar in copper pan, add glucose (corn syrup) and water; set on fire and stir until sugar dissolves and starts to boil, then add cream or condensed milk gradually, and cook to "the ball." Take off the fire, stir a little, then add stiff cream fondant and beat the batch until well mixed, add flavor, set on the fire again a minute and stir until sufficiently warmed to pour out easily. Pour on greased slab to cool, then mark the top and glaze with white of eggs. When dry, break into suitable pieces.

NOTE:

If a very rich quality is wanted, a half pound butter may be added while the batch is cooking.

Source: Savage Brothers Co.

Fudge for Slicing

			% by Wt.
Condensed skimmed or whole milk (40% sugar)	30	lb	11.28
Granulated sugar	50	lb	18.80
Corn syrup	30	lb	11.28
Nulomoline (invert sugar)	10	lb	3.76
Hardened coconut oil (86–110 F m.p.)	10	lb	3.76
Basic Fondant (q.v.)	80	lb	30.08
Frappé No. 2 (q.v.)	10	lb	3.76
Powdered salt	1	lb	0.38
Vanilla or other flavors			to suit
Whole Brazil nuts (medium size)	45	lb	16.90
			Total 100.00

Procedure:

Place the condensed milk, corn syrup, **Nulomoline,** and coconut oil into a kettle and heat and mix until the fat is dissolved, then add the sugar and continue to mix and cook the batch to 250–254 F (at sea level).

Permit the batch to cool to approximately 180 F, add the Brazil nuts, Basic Fondant, frappé, salt, and flavor, and mix thoroughly, then turn the batch out into pans or boxes lined with heavy waxed paper.

When making chocolate flavored fudge, mix from 14–16 lb of cocoa powder with 10 lb of **Nulomoline** and add this to the batch after it has been cooked to 250–254 F, and continue as for vanilla fudge.

BASIC FONDANT:

Granulated sugar	80 lb
Corn syrup	20 lb
Nulomoline (invert sugar)	10 lb
Water (to dissolve the sugar)	

Procedure:

Boil the granulated sugar, corn syrup, **Nulomoline**, plus sufficient water to dissolve the sugar to 244 F. Cool the syrup to 125 F, then beat into fondant.

FRAPPÉ NO. 2:

Corn syrup	50 lb
Nulomoline (invert sugar)	50 lb
Albumen (dissolved in water)	2 lb
Water	4 lb

Procedure:

Boil the corn syrup to 245 F. Add **Nulomoline** and mix until it is dissolved. Promptly transfer this syrup to a marshmallow beater, start the beater, then gradually add the albumen which has previously been dissolved in water. Beat the batch until light.

Source: Crompton & Knowles

Light Vanilla Fudge

		% by Wt.
Granulated sugar	36 lb	27.23
Nulomoline (invert sugar)	3 lb	2.27
Corn syrup	24 lb	18.16

Coconut oil (92 F m.p.)	6	lb	4.54
Water	3	lb	2.27
Sweetened condensed whole milk	24	lb	18.16
Fondant (q.v.)	21	lb	15.89
Frappé No. 2 (q.v.)	15	lb	11.35
Salt	3	oz	0.13
Vanilla flavor			to suit
			Total 100.00

Procedure:

Place the sugar, corn syrup, **Nulomoline**, coconut oil, condensed milk, and water into a kettle. Mix, apply heat and with constant stirring, cook to 244 F. Remove from heat. Add the salt and fondant, mix well, then add the frappé and flavor. Mix and stir until the batch becomes heavy, pour on oiled waxed paper. Nut meats may be worked into surface, if desired.

FONDANT:

Granulated sugar	100	lb
Corn syrup	20	lb
Nulomoline (invert sugar)	10	lb
Water (to dissolve the sugar)	25	lb

Procedure:

Cook to 245 F, cool, and beat into fondant.

FRAPPÉ NO. 2:

Corn syrup	50	lb
Nulomoline (invert sugar)	50	lb
Egg albumen (spray-dried powder)	2	lb

Procedure:

Place half the **Nulomoline** and egg albumen into a marshmallow beater. Mix at slow speed. Meanwhile heat the remaining **Nulomoline** and corn syrup to approximately 200 F. Add to the mixture in the beater. Beat until light.

Source: Crompton & Knowles

Special Vanilla Fudge

			% by Wt.
Fondant creme (q.v.)	20	lb	45.45
BOB SYRUP:			
Granulated sugar	12	lb	27.27
Corn syrup	8	lb	18.18
Water	2	lb	4.55
Salted dairy butter	2	lb	4.55
		Total	100.00

Procedure:

First prepare a rich fondant creme.

Cook bob syrup ingredients altogether to 245 F. Add the bob syrup to fondant creme, mixing well. Permit the batch to set until it is cooled to approximately 175 F, then add salt, vanilla flavor, and nuts, as desired. The batch is now ready to be spread onto oiled waxed paper.

FONDANT CREME:		
Granulated sugar	100	lb
Cream (36–40% butterfat)	16	lb
Unsweetened evaporated milk	8.9	lb
Nulomoline (invert sugar)	16	lb
Water	0	

Procedure:

Place the sugar, cream, milk, and **Nulomoline** into a kettle, mix well, apply heat, and after the batch has boiled for 2 or 3 min, wash all crystals from the kettle and cook to 244 F. Pour the batch at once on a clean fondant beater and immediately sprinkle the surface of the batch with 1 oz of water. When the batch has cooled to approximately 110 F, beat it into fondant.

Source: Crompton & Knowles

Vanilla Walnut Fudge

		% by Wt.
PART 1—FONDANT (q.v.)	10 lb	42.43
PART 2:		
Fine granulated sugar	5 lb	21.22
Corn syrup	2 lb	8.49
Condensed whole milk	4 lb	16.98
Coconut butter (86 F m.p.)	1.5 lb	6.37
Salt	1 oz	0.27
Frappé No. 1 (q.v.)	1 lb	4.24
		Total 100.00

Procedure:

PART 1: Prepare the fondant.

PART 2: Place the condensed milk, corn syrup, and coconut butter into a kettle and heat until the fat is dissolved, then add the sugar and stir and cook the batch until it boils. Wash down all grains of sugar that adhere to the inside of the kettle and boil the batch to 240 F.

Turn off the heat, add the salt, flavor, and frappé, then 10 lb of the fondant (Part 1) and mix thoroughly. Spread the batch out on oiled paper to set.

FONDANT:

Granulated sugar	8.5 lb
Nulomoline (invert sugar)	1.5 lb
Cream (22 or 40%)	3 lb

Procedure:

Place the sugar, **Nulomoline**, and 2 lb of the cream into a kettle. Stir the batch constantly until it boils. Wash down all grains of sugar that adhere to the inside of the kettle, add the remaining cream and boil the batch to 245 F. Pour the batch onto a cream cooler and allow to remain undisturbed until it has cooled to approximately 110 F, the beat into fondant.

FRAPPÉ NO. 1:

Corn syrup	50 lb
Invert sugar	50 lb
Albumen	1 lb
Water	2 lb

Procedure:

Cook corn syrup to 245 F (sea level), shut off the heat and add invert sugar and stir until it is melted. Place the batch into a marshmallow beater, start the beater, then gradually add albumen which has been previously dissolved in water and beat the batch until light.

Source: Crompton & Knowles

Chapter 9

HARD CANDIES

There are four varieties of hard candy—clear, pulled, grained, and filled. Hard candies are combinations of sugar or corn syrup and acids such as citric, malic, or tartaric. Hard candy is a supercooled liquid and though made from the same ingredients as fondant, a soft candy, it differs in having a low moisture content. Typical products are lollipops, lozenges, candy sticks, rock candy, sour balls, kisses, mints, and peppermints.

Almond Brittle

			% by Wt.
Sugar	5	lb	30.38
Glucose (corn syrup)	4	lb	24.31
Almonds, chopped	7	lb	42.53
Butter	6	oz	2.27
Salt	2	Tbl	0.38
Lemon extract	1	Tbl	0.13
			Total 100.00

Procedure:

Cook sugar, glucose (corn syrup), and a little water to 280 F; set off and add butter, salt, and extract, stirring until butter is thoroughly dissolved. Then add nuts and stir quickly through the batch. Scrape out on greased slab.

Now pull the batch out until uniformly thin, turning it over occasionally and folding together in order to prevent too much hardening on the bottom. It is more easily handled if the batch is broken into several sections while doing this.

Source: Savage Brothers Co.

Almond Rock

			% by Wt.
Brown sugar	6	lb	66.66
Almond pieces	3	lb	33.30
Lemon extract (to flavor)	0.15	g	0.04
			Total 100.10

Procedure:

Boil brown sugar to the "crack;" it will take much longer than white sugar. Flavor with a few drops lemon extract, and when brittle, pour the almonds that have been made warm into it, and stir enough to mix. Pour inch deep on slab between iron bars, or into greased pans.

Source: Savage Brothers Co.

Butterscotch Toffee Brittle

			% by Wt.
Fine granulated sugar	18	lb	46.61
Nulomoline (invert sugar)	2	lb	5.18
Dairy butter (unsalted)	16	lb	41.42
Whole almonds	2.5	lb	6.47
Salt	2	oz	0.32
Vanilla flavor			to suit
			Total 100.00

Procedure:

Place the butter in a clean copper kettle and heat until melted. Add the sugar and stir well; then add the **Nulomoline.** Allow the batch to cook on a moderate fire to the boiling point; then wash down the sides of the kettle using the smallest amount of water possible. Continue to stir and boil the batch to approximately 260 F.

Now add the almonds which should be unroasted. Continue to stir the batch until it becomes "livery" in consistency. This is the critical point and the temperature will be in the neighborhood of 290 F. The consistency varies considerably with the quality of the butter used.

After the batch has been cooked to the consistency described above, add the salt and flavor and spread immediately on an oiled slab with a palette knife; level it off and distribute the nuts evenly throughout the batch.

Cut the batch into pieces with a roller knife and break apart when cold; then coat in a combination of milk chocolate mixed with ground roasted almonds, Brazils, etc.

Source: Crompton & Knowles

Cherry Rock

			% by Wt.
Granulated sugar	1	lb	80.00
Candied cherries	4	oz	20.00
			Total 100.00

Procedure:

Allow 4 oz candied cherrries per pound of sugar. Boil the sugar to the "crack." When brittle, but before the syrup changes color at all, drop the cherries (slightly warmed) into it. Pour out into greased pans or on slab between bars about 2 in. thick.

Source: Savage Brothers Co.

Crispies

			% by Wt.
Sugar	85	lb	69.68
Nulomoline (invert sugar)	15	lb	12.29
Water	22	lb	18.03
			Total 100.00

Procedure:

These candies can be cooked on either an open fire or in a Simplex cooker. For open fire cook altogether to 320 F.

For the Simplex cooker, cook altogether to 270 F, then close the vacuum and pull for 5 in from 15 in.

Pour the batch out on cooler, add color and flavor, then pull the batch as light as possible.

Reserve one-quarter of the batch for an outer jacket. Form the balance for an inner jacket and put the center (formula below) in this jacket. Pull out and fold together four times, then put on the outer jacket and the batch will be ready to spin and cut.

This candy is purposely made extra dry and will grain. It should sell well during the summer months. If coated with chocolate shortly after they are made, they will not grain rapidly.

CENTER FOR CRISPIES		
Coconut oil (76 F m.p.)	5	lb
Powdered sugar	10	lb
Citric acid (powdered)	0.25	oz

Procedure:

Heat the coconut oil to 125 F, add the powdered sugar and citric acid, and mix well. Heat to 165 F.

Source: Crompton & Knowles

Hard Candy

			% by Wt.
Clinton corn syrup*	45	lb	45.00
Sucrose	45	lb	45.00
Water	10	lb	10.00
Flavor and color			to suit
			Total 100.00

Procedure:

Combine corn syrup, sucrose, and water in steam kettle. Heat to 270 F. Close kettle and start vacuum pump. Cook until maximum vacuum has been reached (approximately 27 in.). Pour out. Direct to appropriate areas for adding flavors and color.

Spin hot candy. Cut or shape as desired.

Source: ADM

* **Clinlo 1230** or **Clinton** regular corn syrup

Heath-Type Bar

			% by Wt.
Granulated sugar	55	lb	55.00
Salted dairy butter	35	lb	35.00
Bitsyn® or nut pieces	10	lb	10.00
			Total 100.00

Procedure:

Cook sugar and butter with 10% added water over a very low fire to 290 F. Add **Bitsyn®** and finish at 310 F. Pour onto slab, spread thin. Cut to 4 × 1 in. bars. Coat in milk chocolate.

Source: Pillsbury Co.

"It Clicks"

(For Bar, Bulk, or Package Line)

			% by Wt.
Sugar	45	lb	83.33
Nulomoline (invert sugar)	5	lb	9.26
Molasses	4	lb	7.41
Water (to dissolve the sugar)			
			Total 100.00

Procedure:

Boil the sugar, **Nulomoline**, and water to 325 F, gradually add the molasses, mix thoroughly and pour the batch out to cool. Sprinkle 3 oz of powdered salt over the surface, add mint oil and butter flavor, fold the batch together and when partially cooled, pull well on the hook.

Place the batch in front of the batch warmer, cut off one quarter of the batch and knead the air out of this part. Roll the balance of the batch into a sheet to measure approximately 24 in. wide and 36 in. in length. The sheet should be approximately $^{1}/_{2}$ in. thick.

Meanwhile, prepare the center (q.v.). Spread the center which has been heated to 160–175 F over the sheet of candy so as to cover the entire sheet to a thickness of $^{1}/_{2}$ in., taking care not to cover the ends which are to be joined together. Roll the batch much in the same manner as when forming a jelly roll, closing the ends, then stretch and double it 6 or 8 times in order to produce a honeycomb effect.

Place the kneaded section of the batch around the center to form a skin and spin the batch into pieces of the desired size and coat with chocolate.

CENTER

			% by Wt.
Coconut oil (76 F m.p.)	5	lb	25.00
Cocoa powder	5	lb	25.00
Powdered sugar	10	lb	50.00
			Total 100.00

Procedure:

Heat the coconut oil until melted, add the cocoa powder, mix well, then add the powdered sugar and heat the batch to 165–175 F.

Source: Crompton & Knowles

Jacketed Creams

(Grained Hard Candy Jacket Filled with Fondant Cream)

Hard candy jacket may be boiled on the open fire, continuous cooker, Simplex cooker, or old-style vacuum pan, using any standard formula for hard candy.

If a short jacket is preferred, it is advisable to use minimum amounts of corn syrup or, better yet, make an all-sugar hard candy jacket.

OPEN FIRE WORK:

			% by Wt.
Granulated sugar	36	lb	90.00
Nulomoline (invert sugar)	4	lb	10.00
Water (to dissolve the sugar)			
			Total 100.00

Procedure:

Boil the batch to 310—315 F (sea level).

SIMPLEX COOKER:

		% by Wt.
Granulated sugar	45 lb	81.81
Nulomoline (invert sugar)	10 lb	18.19
Water (to dissolve the sugar)		
		Total 100.00

Procedure:

Boil the batch to 270 F. Pull the vacuum for 5 min from 15 in.

NOTE:

This formula should also work satisfactorily in the old-style vacuum pan.

CONTINOUS COOKER:

		% by Wt.
Granulated sugar	65 lb	65.00
Corn syrup	25 lb	25.00
Nulomoline (invert sugar)	10 lb	10.00
Water (to dissolve the sugar)		
		Total 100.00

Procedure:

Boil the batch to 236 F. Draw into the continuous cooker, pumping approximately 10 lb per min, using 80–90 lb steam pressure.

PREPARATION OF FONDANT CENTER:

Heat any standard fondant to 125–135 F. Use not less than 1 lb of fondant to each 2 lb of the hard candy jacket. For example, for a 40 lb batch of hard candy on the open fire, it would be necessary to melt 15 lb of fondant, to which would be added 10 cc **Sucrovert** (about $1^1/_2$ tsp), mix well, then add $1^1/_2$ lb of Frappé No. 1 (q.v.), color and flavor and mix well.

Meanwhile, roll a portion of the clear hard candy to form a thin sheet, then place this into a box or container, thus forming a line. The prepared fondant center is then poured on top of the clear hard candy in the box and the hard candy immediately folded around the center.

The balance of the hard candy batch, which has been pulled, is wrapped around the clear portion containing the center, and the batch is then spun, formed, and cut on a plastic machine, continuous cutter, old fashioned buttercup cutter, or roller knife.

FRAPPÉ NO. 1:

Corn syrup	50	lb
Invert sugar	50	lb
Albumen	1	lb
Water	2	lb

Procedure:

Cook corn syrup to 245 F (sea level), shut off the heat and add invert sugar and stir until it is melted. Place the batch into a marshmallow beater, start the beater, then gradually add albumen which has been previously dissolved in water and beat the batch until light.

SUGGESTED FLAVOR AND COLOR COMBINATIONS:

Jacket: Raspberry flavor, pink color; Center: White center, trace of raspberry, acid to taste

Jacket: Molasses flavor; Center: Green color, mint flavor

Source: Crompton & Knowles

Lemon Rock

			% by Wt.
Sugar	1	lb	94.10
Lemon juice	1	oz	5.90
		Total	100.00

Procedure:

Make the same way as orange rock (q.v., page 265), except that the juice of half a lemon (for each pound of sugar) is squeezed in when the syrup reaches the "crack."

Source: Savage Brothers Co.

Meloripe Nut Crisp

			% by Wt.
Granulated sugar	12	lb	32.43
Corn syrup	7	lb	18.92

Nulomoline (invert sugar)	5	lb	13.52
Water (to dissolve the sugar)			
Sliced Brazil nuts	12	lb	32.43
Nulco MeloBan (powdered banana product)	1	lb	2.70
Red or yellow color			to suit
			Total 100.00

Procedure:

Boil the sugar, corn syrup, **Nulomoline**, and water to 310 F, add the color, then add the Brazil nuts which have been warmed and mixed with the **Nulco MeloBan**. Mix the batch thoroughly, then turn the batch out on an oiled slab and roll it to the required height. Cut with a roller knife into squares, oblongs, or larger bars, or grind the candy into small particles for use in or on ice cream.

Source: Crompton & Knowles

Molasses Chips

			% by Wt.
Sugar	18	lb	81.82
Water (to dissolve the sugar)			
Molasses (**Grandma's**)	4	lb	18.18
Powdered salt			to suit
			Total 100.00

Procedure:

Place the sugar, water, and 2 lb of the molasses into a kettle and stir until the batch starts to boil. Carefully wash down all grains of sugar that adhere to the inside of the kettle and boil the batch to 310 F. Gradually add in a fine stream the remaining 2 lb of molasses, mixing constantly, and when the batch registers a temperature of 310 F (sea level), pour it out to cool.

Sprinkle the powdered salt over the surface, fold together, pull on the hook and spin into chips or pass the pulled batch through drop rolls.

HONEYCOMB:

Have a rolling pin or metal cylinder or pipe warm. Roll the batch down to about $^1/_2$ in. high then place it around the cylinder or rolling pin, taking care that it does not stick. Hold one end and pull out the rolling pin. Force a little air into the opening in the batch, pull the batch out 3 or 4 ft in length, keeping it in shape so as to keep the center hollow. Stretch the batch 4 or 5 times, doubling each time. Repeat this until the desired number of folds are obtained, then spin the batch into pieces of the desired size, cutting on a buttercup cutter, chip cutter, or roller knife.

Source: Crompton & Knowles

Molasses Cream-Filled Sticks

JACKET:

			% by Wt.
Sugar	6	lb	59.63
Molasses (**Grandma's**)	4	lb	39.75
Salt	1	oz	0.62
			Total 100.00

Procedure:

Add sufficient water to dissolve the sugar and cook to 290 F.

FONDANT CENTER:

			% by Wt.
Sugar	8	lb	72.73
Nulomoline (invert sugar)	1	lb	9.09
Corn syrup	1	lb	9.09
Butter	1	lb	9.09
			Total 100.00

Procedure:

Place all of the ingredients in a kettle with sufficient water to dissolve the sugar and cook to 250 F. Then place in the cream beater and cream same as fondant.

Have the fondant center sufficiently warm so that it will not chill the jacket. In the meantime cook the above jacket and when cooled on the cooling table sufficiently to be plastic, roll the fondant center in an oblong roll and wrap the jacket around it. This can be then spun out to about the size of a broom handle and when sufficiently cooled should be marked and broken into the desired lengths.

Powdered sugar should be dusted on the spinning table when this is spun. This will start a grain in the jacket but will prevent it from becoming sticky.

Source: Crompton & Knowles

Molasses–Peanut Butter Puffs

(Simplex Cooker)

JACKET:

		% by Wt.
Sugar	85 lb	85.00
Nulomoline (invert sugar)	15 lb	15.00
		Total 100.00

Procedure:

Cook together in a Simplex cooker. When the temperature of the batch reaches 275 F, reduce the gas or steam, close the vacuum and when the gauge registers 15 in., note the time. From this point the batch should be kept under vacuum for 5 min. As high a vacuum as possible should be secured. Pour out, cool and handle as usual.

Any desired stripe and color combination may be used.

CENTER:

		% by Wt.
Molasses	20 lb	40.00
Peanut butter	30 lb	60.00
		Total 100.00

Procedure:

Cook the molasses to 235 F. Add this to the peanut butter and mix well.

Form the jacket and add the center, taking care to have the center warm enought not to chill the jacket. Spin and cut into any desired shape or size.

Use about 1 lb of center to each $1^1/_2$ lb of jacket.

Source: Crompton & Knowles

Molasses Peanut Crisps

		% by Wt.
Granulated sugar	25 lb	32.79
Corn syrup	10 lb	13.12
Nulomoline (invert sugar)	5 lb	6.55
Molasses (**Grandma's**)	7.5 lb	9.84
Coconut butter	2.5 lb	3.28

Salt	12	oz	0.98
Soda	8	oz	0.65
Peanuts	25	lb	32.79
			Total 100.00

Procedure:

Place into a kettle the sugar, corn syrup, **Nulomoline**, molasses, coconut butter, and salt and stir and cook to 245 F, then add the peanuts. These should be unroasted and chopped fine (granulated). Continue to stir and cook the batch to 290 F, take off the fire, add the soda and a little vanilla flavor. Stir through quickly and dump the batch onto a cooler.

This candy may be finished by being handled as a regular hard candy batch and run through a chip roller, or it can be spread thin and cut with adjustable cutters.

As soon as the candies are cold, coat them with chocolate.

Source: Crompton & Knowles

Orange Rock

			% by Wt.
Sugar	1	lb	71.00
Orange peel	0.4	lb	29.00
			Total 100.00

Procedure:

Allowing 1 cup of candied orange peel, shredded, to a pound of sugar, proceed as in Cherry Rock recipe (q.v.).

Source: Savege Brothers Co.

Peanut Brittle

			% by Wt.
Sugar	40	lb	40.00
Clinton regular corn syrup (43 Bé)	25	lb	25.00
Water	10	lb	10.00
Spanish peanuts, raw	24.38	lb	24.38

Salt		to suit
Bicarbonate of soda	10 oz	0.62
		Total 100.00

Procedure:

Cook sugar, water, and **Clinton** regular corn syrup over open flame to 240 F. Add peanuts and cook to approximately 310 F or to a golden brown color and remove from fire. Add salt to suit and bicarbonate of soda. Stir until soda is mixed and continue stirring for a short while.

Permit batch to foam or "puff up" and pour on oiled slab. Quickly spread to the desired thickness, turn over on slab as soon as the surface is cool enough to handle, and spread or stretch into as thin a layer as possible.

Source: ADM

Peppermint Stick Candy

		% by Wt.
Sugar	15 lb	78.91
Water	2 qt	21.04
Cream tartar	2.5 g	0.03
Peppermint oil	1.5 g	0.02
		Total 100.00

Procedure:

Place sugar in a copper pan, add water and cream tartar. Set pan on a good fire and stir until sugar starts to boil, then cover the pan with a galvanized iron or wood cover, and allow to steam for a few minutes. Now remove the cover, place your thermometer in the batch, and cook to 330 F.

Pour out on greased marble, leaving as much in the pan as is needed for striping, approximately 2 lb, color this red and after warming the pan again, run it out on the marble and as soon as cool, fold it together in a lump and keep it warm before the batch heater.

Now fold up the main batch, add peppermint oil, and pull a little over the hook, not too much, twist out the air and form into a round, oblong piece.

Knead the red piece a little, roll and pull out into thin strips, laying them over the main batch about 2 in. apart, then roll the whole batch a little, warm the end while you pull it out, and then start to spin and twist as per directions for lemon stick. It is the helper's duty to see that the batch is kept in shape and not allowed to flatten out at all.

Source: Savage Brothers Co.

Pure Sugar Hard Candy Lemon Drops

			% by Wt.
Coarse granulated cane sugar	44	lb	67.69
Cold water	15	lb	23.08
Nulomoline (invert sugar)	6	lb	9.23
			Total 100.00

Procedure:

Place all together into a clean copper pan or equivalent and gradually bring the batch to the boiling point, then carefully wash down all grains of sugar that adhere to the sides of the pan.

Place a clean cover over the boiling syrup so that the steam from the batch strikes the sides of the pan and dissolves the particles of sugar that may still cling to the pan.

After the batch has boiled for 3–4 min, remove the cover and boil the batch as rapidly as possible to 320–330 F (according to desired color).

Without delay pour the batch on a cold water cooled slab, allow the batch to remain undisturbed for 2 or 3 min, then turn in the chilled edges of the batch. Now take a part of the partially cooled batch and stretch it as thin as tissue paper and hold it up to the light. Small, almost imperceptible grains of sugar should appear in the candy, possibly 6 grains to the inch or thereabouts.

For candies that are not to be pulled (clear work) it is advisable to doctor the batch so that there is definite evidence of grain for the partially chilled batch, as described immediately above. But for gloss work, it is desirable that the batch be almost free from grain; in other words, some grain is desirable for gloss work, but not as much grain as should show up in clear work.

After the batch is partly cooled, fold it together and add a little yellow color, working it into the hot part of the batch. Meanwhile mix together from 10 to 14 oz of finely powdered citric acid with sufficient lemon oil, possibly $^{1}/_{2}$ oz, then work this into the batch just enough to evenly distribute the flavoring materials. Avoid working the batch too much unless opaque candy is desired.

SPECIAL SUGGESTIONS:

After the batch has been brought to the boiling point, it should be boiled as rapidly as possible to the desired temperature so as to secure uniform workmanship and color.

Pour the batch out as rapidly as possible; particularly avoid allowing the batch to remain on the hot fire after it has reached the proper temperature.

Do not scrape any part of the batch from the kettle onto the slab as this may cause the entire batch to grain.

Allow the batch to drain from the pan onto the cooling slab, taking special care not to allow an undue amount of the boiled batch to remain in the cooking pan.

It is not necessary to wash out the pan after each boiling but more uniform work will be secured if this is done.

Allow plenty of cooling surface on which to chill the batch, for if a batch is poured on a warm slab the batch may become discolored.

Scrap from each succeeding batch may be worked into a new boiling by placing the scrap on the properly oiled slab, and pouring the hot batch on top.

It is decidedly important that the cooking time per batch be carefully regulated so as to cook each batch in a known period; otherwise the color and quality of the candy will not be identical.

If more than one fire is used, be certain the cooking timer per batch will be the same on each fire, otherwise it may be necessary to use more **Nulomoline** in a batch that cooks more rapidly than a batch cooked on a slower fire and vice versa.

For gloss work it may be necessary to increase the amount of **Nulomoline** as per formula for lemon drops, but this can be determined after a few boilings.

All hard candy if exposed to damp air will eventually become sticky, therefore, after the candies have been run through the drop rolls or spun, they should be cooled and as soon as cold they should be packed in airtight containers. The container must be absolutely free from moisture.

All hard candy will either grain or become sticky, regardless of how it is made. By using **Nulomoline** it is possible to scientifically doctor the batch, and thus strike a happy medium, and therefore more effectively control the appearance, flavor, color, and texture, particularly the keeping qualities of the hard candies.

"**Nulomoline**-made" hard candies are more easily worked or spun, because it gives a certain flexibility to the batch that causes a reduction in scrap. **Nulomoline** is a dependable "doctoring" agent.

Source: Crompton & Knowles

Pure Sugar Hard Candy

(Made in Old-Style Closed-Type Vacuum)

			% by Wt.
High grade cane sugar	240	lb	64.86
Nulomoline (invert sugar)	60	lb	16.22
Water	70	lb	18.92
		Total	100.00

Procedure:

Place the above ingredients into a melting kettle and after the batch has boiled for a minute or two, carefully wash down the sides of the kettle so as to remove all grains of sugar that might adhere to the sides. Boil the batch to 235 F, using approximately 80 lb of steam pressure, then draw the batch into the vacuum pan, maintaining a constant steam pressure of 80 lb and gradually increase the vacuum to 10 in.

Hold the vacuum steadily at 10 in. until the batch reaches a temperature of approximately 275–280 F (according to the color desired).

After the thermometer registers a temperature of 275–280 F, reduce the steam pressure to the minimum and increase the vacuum from 24 to 29 in. After the vacuum gauge indicates a vacuum of 25 in. or more, allow the vacuum to pull for 5–7 min.

The batch may then be removed from the vacuum pan and handled in the usual manner.

Source: Crompton & Knowles

Chapter 10

MARSHMALLOWS

Cast-Grained Marshmallow

			% by Wt.
Granulated sugar	65	lb	58.89
Water (to dissolve the sugar)			
Corn syrup	25	lb	22.65
Nulomoline (invert sugar)	10	lb	9.06
Gelatin (medium grade soaked in 8 lb water)	2.25	lb	2.04
Water	8	lb	7.25
Sucrovert (invertase)	2	oz	0.11
			Total 100.00

Procedure:

Boil the sugar and water to 236 F. Turn off the heat, add the corn syrup and **Nulomoline**, then add the gelatin which has been soaked in the water for at least one-half hour and mix until the gelatin is completely dissolved.

Place the batch at once into a marshmallow beater and beat until quite light, then add the **Sucrovert**, color, and flavor, and cast the marshmallow into starch which has been tempered to 90–100 F.

Allow the marshmallows to remain in the starch in a room of normal temperature for several hours to permit the gelatin to set, then transfer the candies to a drying room tempered to 105–115 F. Allow the marshmallows to remain in the drying room for approximately 8 h, after which they may be removed from the starch.

Source: Crompton & Knowles

Cast Molasses Marshmallow

			% by Wt.
Water	28	lb	21.46
Good gelatin	2.5	lb	1.92
Granulated sugar	40	lb	30.65
Molasses	15	lb	11.49
Nulomoline (invert sugar)	10	lb	7.66
Corn syrup	35	lb	26.82
			Total 100.00

Procedure:

Mix the gelatin with 35% of the water and set aside for at least 15 min.

Place the remaining 65% of water into a kettle, add the sugar and stir and heat until the sugar is dissolved. Turn off the heat, add the soaked gelatin and mix until the gelatin is dissolved. Add the molasses, **Nulomoline**, and corn syrup.

Place the batch into a marshmallow beater and beat until the marshmallow attains a volume of 1 lb to 1 qt.

Add mint flavor, mix well, then cast the batch into the starch. Allow the marshmallows to set overnight, then remove from the starch, coat them with chocolate icing and roll in chocolate shot.

Source: Crompton & Knowles

Cast Tender Short Marshmallow

			% by Wt.
Corn syrup	50	lb	30.37
Sugar	25	lb	15.19
Nulomoline (invert sugar)	25	lb	15.19
Starch (#50–60)	7	lb	4.25
Water	56	lb	34.02
Cream of tartar	2	oz	0.08
Albumen (dissolved in 2 lb of water)	1	lb	0.60
Gelatin (medium grade, dissolved in 1.5 lb of water)	8	oz	0.30
			Total 100.00

Procedure:

Boil the sugar, corn syrup, **Nulomoline**, starch, water, and cream of tartar to a good thread (22 F). Place the batch into a marshmallow beater, gradually add the albumen and gelatin solutions and beat until light.

Cast the batch into starch molds, either plain, or onto nutmeats or cut fruit.

Nulco MeloBan may be added on the basis of 4 lb to the above sized batch.

Place the marshmallow-filled starch trays into a drying room (temperature 130 F) for approximately 15 h or until the marshmallow is sufficiently firm to remove from the starch.

Coat the marshmallows with jelly, icing, caramel, or chocolate.

Source: Crompton & Knowles

Cast White Marshmallow

(For Packaging, No Boil Process)

			% by Wt.
Gelatin	1.84	lb	1.84
Water	17.38	lb	17.38
Granulated sugar	26.75	lb	26.75
Clintose brand dextrose	13.37	lb	13.37
Water	13.91	lb	13.91
Clinton regular corn syrup (43 Bé)	26.75	lb	26.75
Vanilla flavoring			to suit
			Total 100.00

Procedure:

Dissolve gelatin in water previously heated to 150–160 F. Add sugar and dextrose mixed with remaining water. Add corn syrup and beat the marshmallow to 3.5 lb/gal, or to the desired consistency. Add vanilla flavoring near the end of mixing.

Cast into dry clean starch (100 F). The moisture content of the starch should be 5–6%. Let stand in starch overnight.

Shake out on the following day and dust with a mixture of 50% powdered sugar/50% starch. This mixture should be previously dried out in a hot room or drying cabinet.

Let marshmallow set for several hours before packing.

Source: ADM

Chewy Grained Marshmallow

			% by Wt.
Gelatin (225 bloom)	2.75	lb	1.43
Granulated sugar	80	lb	41.49
Nulomoline (invert sugar)	15	lb	7.78
Corn syrup	40	lb	20.75
Water	35	lb	18.15
Extra X powdered sugar	20	lb	10.37
Sucrovert (invertase)	1	oz	0.03
Flavor and color			to suit
			Total 100.00

Procedure:

Mix the gelatin with half of the water (either warm or cold). Place the remaining amount of water into a kettle, add the sugar, mix and heat to approximately 200 F. Add the soaked gelatin, mixing until the gelatin is completely dissolved.

Add the **Nulomoline**, corn syrup, powdered sugar, and **Sucrovert** and beat the batch until 1 gal will weigh approximately 5 lb net. Add color and flavor, mixing thoroughly.

At this point the batch will register a temperature of 110–115 F.

Cast into starch at normal temperature, allowing the marshmallow to set for several hours before transferring to a drying room tempered to 105–110 F where it is permitted to remain for 5–6 h, after which the grained marshmallow is removed from the starch and is ready to be packed into cellophane bags; or the grained marshmallow may be cast into squares, oblongs, or other shapes and later coated with icing or chocolate.

This formula will produce a chewy grained-type marshmallow, but if a shorter, less flexible marshmallow is desired use 5 oz less of the gelatin.

Source: Crompton & Knowles

Chocolate-Coated Cut Marshmallow

			% by Wt.
Granulated sugar	90	lb	35.57
Nulomoline (invert sugar)	30	lb	11.86
Corn syrup	80	lb	31.62

Good grade gelatin	5	lb	1.98
Water	48	lb	18.97
Vanilla flavor			to suit
			Total 100.00

Procedure:

Soak the gleatin in 20 lb of the water for at least 15 min.

Meanwhile, heat the sugar with the balance of the water until the sugar is dissolved. (Do not permit the batch to boil.) Now add the **Nulomoline** and soaked gelatin and mix thoroughly until the gelatin is dissolved. Add the corn syrup and beat the batch until light. Add the flavor, mix well, and spread the batch out on paper or oiled trays, and when it has set, cut into bars and coat with chocolate.

Source: Crompton & Knowles

Cut Marshmallow

No. 1

			% by Wt.
Fine granulated sugar	55	lb	38.52
Nulomoline (invert sugar)	15	lb	10.51
Corn syrup	30	lb	21.02
High grade gelatin	2.75	lb	1.93
Cold water	40	lb	28.02
			Total 100.00

Procedure:

Mix the gelatin with the water, then place this into a jacketed steam kettle. Turn the steam on partly and mix continuously until the gelatin is completely dissolved. The batch at this point will register a temperature of approximately 145 F.

Now add the granulated sugar and continue to heat the batch until the sugar is entirely dissolved. Turn off the heat, add the corn syrup and **Nulomoline**. Mix thoroughly and the batch is now ready to beat.

Place the batch into a marshmallow beater and allow it to beat until one American gallon measure filled with this marshmallow will weigh approximately 4 lb net.

Add flavor as desired, mix well and spread out on tables or boards lined with ordinary wrapping paper. After the marshmallow has set, turn the marshmallow sheets upside down, moisten the paper so as to remove it from the marshmallow and the marshmallow sheets are now ready to be cut.

Sprinkle either granulated sugar or toasted coconut over the marshmallow pieces.

NOTES:

If toasted coconut is used, in order to keep it tender and prevent the coconut from absorbing moisture from the marshmallow, it is suggested that the coconut be toasted as follows:

Place into a jacketed steam kettle equipped with a good mixing device 10 lb of **Nulomoline** and allow to heat until the **Nulomoline** is melted, then add 50 lb of unsweetened desiccated coconut. Allow the coconut to slowly roast mixing it continuously and when a light brown color has been developed, turn off the heat and spread the coconut out on cooling slabs.

It is important not to roast the coconut too dark a color as it will continue to become darker in color while it is being cooled.

Source: Crompton & Knowles

No. 2

(For Dipping in Chocolate or Rolling in Toasted Coconut, Etc.)

			% by Wt.
Fine granulated sugar	50	lb	33.84
Corn syrup	40	lb	27.07
Nulomoline (invert sugar)	20	lb	13.54
High grade gelatin	2.75	lb	1.86
Cold water	35	lb	23.69
		Total	100.00

Procedure:

Mix the gelatin with water, then slowly heat until it is dissolved. Now add the sugar, increase the heat and continue to stir the batch until the sugar is completely dissolved. The batch at this point will register a temperature of approximately 175 F. Turn off the heat, add the **Nulomoline** and corn syrup and stir until the **Nulomoline** is dissolved. Beat the batch at medium speed until it begins to swell. Continue to beat the batch until one American gallon filled with marshmallow will weigh approximately 3$^3/_4$–4 lb net. Color and add flavor. Spread into trays lined with ordinary wrapping paper.

After the marshmallow has set in a room of normal temperature for approximately 5–10 h, the paper should be moistened with water and removed from the marshmallow. The marshmallow may then be cut into squares or bars by the use of a moistened roller knife. The marshmallows should be allowed to dry sufficiently, before dipping in chocolate or rolling in coconut or colored sugar.

NOTE:

For each 550 ft above sea level, reduce the boiling temperature by 1 F.

Source: Crompton & Knowles

No. 3

		% by Wt.
Gelatin (#225 test)	2.75 lb	2.07
Water	20 lb	15.07
Granulated sugar	50 lb	37.66
Nulomoline (invert sugar)	20 lb	15.07
Corn syrup	40 lb	30.13
Color and flavor		to suit
		Total 100.00

Procedure:

Mix gelatin with half of the water, either warm or cold, and set aside for at least 15 min. Place the remainder of the waer into a pan, add the sugar and **Nulomoline.** Heat until the sugar is dissolved. Do not boil. Then add the gelatin–water mixture, mixing well, until the gelatin is dissolved. Then add the corn syrup. Place the batch into an upright marshmallow beater and beat until 1 gal of the marshmallow will weigh approximately 4 lb net. Color and flavor may now be added, and the batch spread onto ordinary wrapping paper to form a sheet of the required height. Sprinkle the surface of the batch with desiccated macaroon coconut. When the batch has set, it may be cut into pieces of the desired size and shape, then rolled in desiccated coconut. Permit the cut marshmallow to dry on the surface before packing closely together.

SUGGESTION:

Spread half the batch on the wrapping paper. Sprinkle chocolate bits on surface and then spread remaining half of batch on top proceeding as above.

Source: Crompton & Knowles

Cut Flavored Marshmallow

		% by Wt.
Gelatin (#225 bloom test)	2.75 lb	2.07
Water	20 lb	15.07
Granulated sugar	50 lb	37.66
Corn syrup	40 lb	30.13
Nulomoline (invert sugar)	20 lb	15.07
Color and flavor		to suit
Garnish with colored sprinkles		
		Total 100.00

Procedure:

Mix gelatin with half of the water, either warm or cold, and set aside for at least 15 min. Place the remainder of the water into a pan, add the sugar and heat until the sugar is dissolved. Do not boil. Then add the gelatin–water mixture, mixing well, until the gelatin is dissolved. Then add the corn syrup and **Nulomoline** and place the batch into an upright marshmallow beater and beat until 1 gal of the marshmallow will weigh approximately 4 lb net. Color and flavor may now be added, and the batch spread onto ordinary wrapping paper to form a sheet of the required height. Sprinkle the surface with colored decorettes, then sprinkle the surface of the batch with desiccated macaroon coconut. When the batch has set, it may be cut into pieces of the desired size and shape, then rolled in desiccated coconut. Permit the cut marshmallow to dry on the surface before packing closely together.

SUGGESTION:

Spread half the batch on the wrapping paper. Sprinkle chocolate bits on surface and then spread remaining half of batch on top proceeding as above.

Source: Crompton & Knowles

Cut Marshmallow Squares and Bars

(Rolled in Toasted Coconut)

			% by Wt.
Granulated sugar	25	lb	20.75
Nulomoline (invert sugar)	15	lb	12.45
Corn syrup	60	lb	49.79
Water	18	lb	14.94
Good grade gelatin (#200)	2.5	lb	2.07
			Total 100.00

Procedure:

Soak the gelatin for at least 15 min in 10 lb of the cold water.

Meanwhile, dissolve the sugar in the remaining 8 lb of water. Do not permit the batch to boil. Turn off the heat, add the **Nulomoline** and soaked gelatin and stir until the gelatin is completely dissolved, then add the corn syrup. Place the batch into a beater and beat until quite light. Add flavor and color as desired and spread the batch out on paper-lined or greased trays and immediately sprinkle the surface with toasted coconut.

When the marshmallow has set, cut it into square or bars and roll in toasted coconut. Permit the bars to dry for several hours before packing.

Source: Crompton & Knowles

Easter Eggs

No. 1

			% by Wt.
Corn syrup	70	lb	57.97
Nulomoline (invert sugar)	30	lb	24.84
Gelatin	2.75	lb	2.28
Water	18	lb	14.91
			Total 100.00

Procedure:

Place the corn syrup into a kettle and cook to 244 F. Turn off the heat, add the **Nulomoline** and stir until melted. In the meantime, have the gelatin soaked in the water and add it to the batch as soon as the **Nulomoline** is melted.

When this is done, place the batch into a beater and beat until 1 qt weighs 1 lb or less.

Source: Crompton & Knowles

No. 2

			% by Wt.
Corn syrup	50	lb	40.64
Sugar	25	lb	20.33
Nulomoline (invert sugar)	25	lb	20.33
Good grade gelatin	3	lb	2.44
Water	20	lb	16.26
			Total 100.00

Procedure:

Place the water and gelatin into a kettle and stir and heat slowly until the gelatin is melted. Add the sugar and continue to heat and stir until the sugar is dissolved. Turn off the heat, add the corn syrup and **Nulomoline** and stir until the **Nulomoline** is melted. Place into a beater and beat until 1 qt weighs 1 lb or less.

Source: Crompton & Knowles

No. 3

			% by Wt.
Nulomoline (invert sugar)	20	lb	15.53
Sugar	40	lb	31.07
Corn syrup	40	lb	31.07
Good grade gelatin	2.75	lb	2.14
Water	26	lb	20.19
			Total 100.00

Procedure:

Place the water and gelatin into a kettle and stir and heat slowly until the gelatin is melted. Add the sugar and continue to heat and stir until the sugar is dissolved. Turn off the heat, add the corn syrup and **Nulomoline** and stir until the **Nulomoline** is melted. Place into a beater and beat the batch until 1 qt weighs 1 lb or less.

Source: Crompton & Knowles

Ever-Soft Marshmallow

			% by Wt.
Nulomoline (invert sugar)	50	lb	46.73
Corn syrup	50	lb	46.73
Gelatin (200 bloom)	2	lb	1.87
Water	5	lb	4.67
			Total 100.00

Procedure:

Soak the gelatin in the water. Heat the **Nulomoline** and corn syrup to approximately 200 F, then add the soaked gelatin and stir until it is dissolved. Place the batch into a marshmallow beater and beat until light, then cast.

If the weather or workroom is damp, it is advisable to sift some starch over the top of the trays. Leave the goods in the starch until the following day, then remove them and coat with chocolate.

Source: Crompton & Knowles

Grained Marshmallow (Peanuts)

			% by Wt.
Granulated sugar	100	lb	49.28
Nulomoline (invert sugar)	10	lb	4.93
Corn syrup	50	lb	24.65
Gelatin (125-g test)	2.75	lb	1.36
Water	40	lb	19.72
Sucrovert (liquid invertase)	2	oz	0.06
Color and flavor			to suit
			Total 100.00

Procedure:

Heat the sugar, **Nulomoline**, and 24 lb of the water to 190–195 F. Meanwhile, add the gelatin which has been melted in 8 lb of hot water, mixing thoroughly, then add the corn syrup and place the batch into a beater starting the beater and beating until light. Add the **Sucrovert** and an additional 8 lb of water to adjust the batch to the desired casting consistency, then add the color and flavor and 2 lb of powdered sugar, mix thoroughly and cast the marshmallow into starch impressions. The marshmallows are permitted to remain in starch impressions at normal room temperature until the gelatin has set, after which the marshmallows are transferred to a drying room tempered to approximately 105–115 F. When the marshmallows have grained, they are then removed from the starch impressions.

Source: Crompton & Knowles

Marshmallow/Bitsyn® Bar

			% by Wt.
PART 1:			
Granulated sugar	620	g	44.83
Water	200	g	14.46
Regular corn syrup	370	g	26.75
Invert sugar	50	g	3.62
PART 2:			
Gelatin (**Swift's** 250° bloom)	23	g	1.66
Water	120	g	8.68
			Total 100.00

Procedure:

Cook Part 1 to 236 F, then pour into bowl of planetary mixer. Add Part 2 to Part 1. Using beater arm and top speed, whip until specific gravity of 0.6148 is reached. (This should still be fluid enough for depositing.)

Deposit into starch in bar shape 2 × 1 × 3/4 in. and leave overnight. Remove from starch and using milk chocolate, either (1) completely coat bar with chocolate, then roll in **Bitsyn®** or (2) pass bars over a prebottomer (coating just one surface), turn bars over and sprinkle **Bitsyn®** on top.

Completely coat each bar in milk chocolate.

Source: Pillsbury Co.

Marshmallow Soufflé

		% by Wt.
Agar agar	1.75 lb	0.95
Water	80 lb	43.66
Granulated sugar	50 lb	27.29
Corn syrup	25 lb	13.64
Nulomoline (invert sugar)	25 lb	13.64
Soy albumen	1.5 lb	0.82
Green color, peppermint oil		to suit
		Total 100.00

Procedure:

Soak the agar agar in all of the water for at least 1 h, preferably longer.

Place the soaked agar agar into a kettle; gradually bring to the boiling point. Add the sugar, corn syrup, and **Nulomoline** and cook the batch to 222 F (sea level). Allow the batch to cool to approximately 180 F.

Place the batch into a marshmallow beater, add the soy albumen (dry) and beat until light. Add the color and flavoring and mix well.

Spread the batch on heavy wrapping paper, leveling off to the required height in much the same manner as when handling cut marshmallow.

Sprinkle granulated sugar or desiccated coconut over the surface and when the marshmallow sets or firms, cut into pieces of the desired size, rolling well in the sugar or coconut. Allow to dry on the surface, then pack.

Source: Crompton & Knowles

MeloBan Cut Marshmallow Squares and Bars

(Rolled in Toasted Coconut)

		% by Wt.
Granulated sugar	25 lb	20.16
Water	8 lb	6.45
Nulomoline (invert sugar)	15 lb	12.10
Corn syrup	60 lb	48.39
Good grade gelatin (soaked in the cold water)	2.5 lb	2.02
Cold water	10 lb	8.06
Nulco MeloBan (powdered ripe banana)	3.5 lb	2.82
Banana color		to suit
		Total 100.00

Procedure:

Soak the gelatin in the 10 lb of cold water for at least 15 min.

Meanwhile, heat the sugar with the 8 lb of water until the sugar is dissolved. (Do not permit the batch to boil.) Turn off the heat, add the **Nulomoline** and gelatin solution and stir until the gelatin is completely dissolved, then add the corn syrup.

Place the batch into a beater and beat until quite light. Sift in the **Nulco MeloBan**, add the banana color and mix thoroughly. Spread the batch out on paper-lined or greased trays and immediately sprinkle the surface with toasted coconut, etc.

When the marshmallow has set, cut it into squares or bars and roll in toasted coconut. Permit the bars to dry for several hours before packing.

Source: Crompton & Knowles

Marshmallows

		% by Wt.
Granulated sugar	7.5 lb	42.44
Gelatin	8 oz	2.83
Glucose (corn syrup)	5.5 lb	31.13
Warm water	0.5 gal	23.60
	Total	100.00

Procedure:

Dissolve the gelatin in water, then soak all together in lukewarm water and beat thoroughly. Beating is one of the main essentials in marshmallow work, and really requires a good mixer or beater.

The batch can now be run out on the slab, marked off and cut into regulation squares, or it can be run in starch if you have the boards and molds, and an extra large mouth funnel dropper.

The batch should be strained before beginning to beat, and if desired may be flavored with orange flower or rose.

Source: Savage Brothers Co.

Plain Cast Flexible Marshmallows

(Using Liquid Sucrose–Sugar Syrup)

			% by Wt.
Liquid sugar	90	lb	67.38
Gelatin (#225)	2 lb 9	oz	1.92
Corn syrup	30	lb	22.46
Nulomoline (invert sugar)	10	lb	7.49
Water	1	lb	0.75
			Total 100.00

Procedure:

Add the water to the liquid sugar, mix in the gelatin and allow it to set for approximately 1/2 h. Then heat the syrup–gelatin mixture to approximately 175 F to completely dissolve the gelatin.

Add the **Nulomoline** and corn syrup, mix well, then place the batch into a marshmallow beater and beat until 1 qt of the marshmallow will weigh approximately 1 lb 15 oz. Add flavor and color as desired and cast the marshmallow into starch impressions.

Source: Crompton & Knowles

Slab Marshmallow

			% by Wt.
Sugar	60	lb	27.52
Dextrose	20	lb	9.17
Water	33	lb	15.14
Gelatin	3.5	lb	1.61
Water	21	lb	9.63
Corn syrup	80	lb	36.70
Vanilla extract	8	oz	0.23
			Total 100.00

Procedure:

Place sugar, dextrose, and first amount of water in kettle and cook to 236–238 F.

Add gelatin that has soaked in water for at least 1/2 h to the syrup. Stir until gelatin has dissolved.

Add corn syrup and place in marshmallow beater. Beat until batch is light and fluffy, but still retains enough heat to spread easily.

Add vanilla extract and pour on slab lined with heavy Kraft paper. Let set overnight before cutting.

SUGGESTIONS:

If too moist, reduce water used to soak gelatin 5 or 6 lb. Dust top surface at once with **Golden Toasted Macaroon Coconut**.

Source: Kraft General Foods Corp.

Chapter 11

NOUGATS

Caramel Nougat Rolls

PART 1—CARAMEL BATCH:		% by Wt.
Hardened edible oil (92–100 F m.p.)	10 lb	4.99
Corn syrup	50 lb	24.94
Granulated sugar	30 lb	14.96
Nulomoline (invert sugar)	20 lb	9.98
Sweetened condensed whole milk	90 lb	44.88
Salt	8 oz	0.25
Vanilla or other flavor		to suit
		Total 100.00

Procedure:

Place into a caramel cooking kettle equipped with double-action agitators all of the above ingredients; heat and cook to a medium firm ball.

The batch may then be split into one or more portions: vanilla flavor may be added to one portion, red color and raspberry flavor added to the second portion, green color and pistachio flavor added to the third portion, and chocolate liquor or cocoa powder may be added to another portion.

The caramel is spread out to form a sheet approximately $^1/_4$ in. thick and should be approximately 15 in. in width so that it may form a definite whirl or jelly-roll effect. The nougat center is then prepared as follows:

PART 2—NOUGAT CENTER:

			% by Wt.
PART 1:			
Egg albumen	2	lb	0.88
Water	4	lb	1.75
Corn starch	2	lb	0.88
Nulomoline (invert sugar)	20	lb	8.77
PART 2:			
Granulated sugar	80	lb	35.09
Corn syrup	100	lb	43.86
Water	15	lb	6.58
PART 3:			
Pink color, vanilla flavor			to suit
Hardened edible oil	5	lb	2.19
			Total 100.00

Procedure:

PART 1: Dissolve the egg albumen in the water, add this to the **Nulomoline** (cold), then add the starch and beat altogether until moderately light.

PART 2: Cook the sugar, corn syrup, and water to 270 F (sea level), pour this into the whipped portion (Part 1), mixing well.

Add the color and flavor and 5 lb of hardened edible oil, mixing just long enough to distribute the oil throughout the batch.

Spread the batch in a thin layer, about $^1/_2$ in. thick over the surface of the cooled sheets of caramel. When the nougat has cooled to approximately 115 F, roll it to produce a jelly-roll effect, then pull, spin, and cut.

This batch may be fed into a standard kiss-wrapping machine and the candies wrapped in either waxed paper or cellophane.

Source: Crompton & Knowles

Cast Molasses Short Nougat for Bars

		% by Wt.
PART 1—FRAPPÉ (q.v.)	35 lb	22.58
PART 2:		
Sugar	50 lb	32.26
Corn syrup	35 lb	22.58
Macaroon	25 lb	16.13
Fondant (q.v.)	10 lb	6.45
Rum and lemon flavor		to suit
		Total 100.00

Procedure:

Boil Part 2 altogether to 255–260 F and pour this at once into the Frappé (Part 1) and mix well. Then add the fondant and mix thoroughly. Add 10 lb of macaroon coconut and a trace of rum and lemon flavor, and again mix well. The batch is now ready to cast into starch.

FRAPPÉ:

Corn syrup	50 lb
Nulomoline (invert sugar)	50 lb
Alubumen (dissolved in water)	3 lb
Water	5 lb

Procedure:

Boil the corn syrup to 245 F. Add the **Nulomoline** and stir until it is melted. Place the batch into a marshmallow beater, start the beater, then gradually add the albumen solution and beat until the batch is quite light.

FONDANT:

Sugar	80 lb
Corn syrup	20 lb
Nulomoline	10 lb

Procedure:

Cook to 242 F.

Source: Crompton & Knowles

Cast or Cut Semichewing Nougat

		% by Wt.
Corn syrup	100 lb	61.19
Granulated sugar	75 lb	
Water (to dissolve sugar)		
Frappé No. $2^1/_2$ (q.v.)	60 lb	20.98
Basic Fondant (q.v.)	50 lb	17.48
Powdered salt	1 lb	0.35
Vanilla flavor		to suit
		Total 100.00

Procedure:

Boil first three ingredients altogether to 260 F (at sea level). Add at once 60 lb of Frappe No. $2^1/_2$, mixing rapidly while adding the frappé.

Now add basic fondant and powdered salt, plus sufficient vanilla flavor and mix thoroughly, then cast the batch into starch impressions.

If a shorter type of nougat is desired, use 90 lb of granulated sugar and 90 lb of corn syrup instead of the above combination.

If chewy nougat is desired, then use 125 lb of corn syrup and 50 lb of granulated sugar, boiling the batch to 258 F.

FRAPPÉ No. $2^1/_2$:

Corn syrup	50 lb
Nulomoline (invert sugar)	50 lb
Albumen (dissolved in water)	2.5 lb
Water	5 lb

Procedure:

Boil the corn syrup to 245 F. Add the **Nulomoline** and stir until it is melted. Promptly place the syrup into a marshmallow beater, start the beater, then gradually add the albumen solution and beat the batch until light.

BASIC FONDANT:

Granulated sugar	80 lb
Corn syrup	20 lb
Nulomoline (invert sugar)	10 lb
Water (to dissolve the sugar)	

Procedure:

Boil the granulated sugar, corn syrup, **Nulomoline**, plus sufficient water to dissolve the sugar to 244 F, cool the syrup to 125 F, then beat the batch into fondant.

Source: Crompton & Knowles

Chewy Nougat (White Color)

			% by Wt.
Egg albumen	46	g	0.81
Vegetable protein	16	g	0.28
Water	125	cc	2.09
Corn syrup	6 lb 5	oz	50.36
Sucrose	5 lb 1	oz	40.39
Vegetable fat	5.4	oz	2.69
Salt	22	g	0.39
Nuts	6	oz	2.99
			Total 100.00

Procedure:

Weigh out egg albumen, vegetable protein, and water. Dissolve protein and albumen in the water. Weigh 1 lb 5 oz corn syrup into beater bowl. Add protein solution to corn syrup. Mix on low speed for 3 min. Weigh into cooking kettle the bob syrup: 5 lb corn syrup, sucrose, salt, and add 1 lb 6 oz water. Cook to 252 F. Remove from fire and cool to 240 F. When cook reaches 252 F run beater with cold mix in it at high speed for 6 min or maximum volume. Melt the vegetable fat. Cool if necessary to maximum temperature of 140 F. Run beater at medium speed and add cooked bob slowly (2 min to add to beater). Mix 1 min after all cooked portion is added. Slow beater to slowest speed and add fat, flavoring, and nuts. Mix very slowly only to complete incorporation of fat. Slab mix on oiled slab and room cool.

Source: Tate & Lyle PLC

Cut Grained Nougat

(For Caramel Coated Pecan Rolls)

			% by Wt.
PART 1—FRAPPÉ (q.v.)	50	lb	32.58
PART 2:			
Granulated sugar	60	lb	39.09
Corn syrup	30	lb	19.54
Water (to dissolve the sugar)			
Casting fondant	10	lb	6.51
Salt	8	oz	0.33
Hardened edible oil	3	lb	1.95
Flavor			to suit
Nut meats, preserved fruit			to suit
			Total 100.00

Procedure:

Boil the sugar, corn syrup, and water to 265–270 F (sea level). Pour this into the frappé (Part 1), beating continuously. Add the fondant, mixing until the fondant is melted, then add the salt.

Beat until the batch shows evidence of graining, then add the edible oil which has been cut into small pieces, and the flavor and then mix well. Now add the nut meats, preserved fruit, and mix to thoroughly distribute, then spread the batch to form sheets.

FRAPPÉ:

Corn syrup	25	lb
Nulomoline (invert sugar)	25	lb
Albumen (dissolved in the water)	1.75	lb
Water	4	lb

Procedure:

Cook the corn syrup to 245 F. Turn off the heat, add the **Nulomoline**, mixing until melted. Place the batch into a marshmallow beater, start the beater, then gradually add the dissolved albumen and beat until light.

Source: Crompton & Knowles

Cut Nougat Taffy

(Chewing)

			% by Wt.
PART 1—FRAPPÉ:			
Corn syrup	20	lb	15.92
Molasses (**Grandma's**)	10	lb	7.96
Nulomoline (invert sugar)	20	lb	15.92
Albumen (dissolved in the water)	2.5	lb	1.99
Water	5	lb	3.98
PART 2:			
Granulated sugar	35	lb	27.86
Corn syrup	15	lb	11.94
Water	8	lb	6.37
Sweetened condensed milk	10	lb	7.96
Powdered salt	2	oz	0.10
Cinnamon and maple flavor			to suit
Pecan pieces			to suit
			Total 100.00

Procedure:

PART 1: Cook the corn syrup and molasses to 250 F, add the **Nulomoline** and mix until the **Nulomoline** is melted. Place the batch into a marshmallow beater, start the beater, then gradually add the dissolved albumen and beat the batch until quite light.

PART 2: Boil the sugar, corn syrup, and water to 290 F, gradually add the condensed milk, mix well, add the salt. Gradually pour the batch into the frappé (Part 1), mixing thoroughly. Add the flavor and nut meats and spread the candy out on oiled slabs to cool. When cool, cut into bars of the desired size and coat with chocolate.

Source: Crompton & Knowles

Divinity Bars and Kisses

(Short Nougat Type, Cut or Rolled)

			% by Wt.
PART 1:			
Albumen (dissolved in the water)	1	lb	2.44
Water	2	lb	4.88
Nulomoline (invert sugar)	5	lb	12.20
PART 2:			
Sugar	25	lb	60.96
Corn syrup	5	lb	12.20
Water (to dissolve the sugar)			
Basic fondant	2	lb	4.88
Nulco MeloBan (powdered ripe banana)	1	lb	2.44
Other flavors, color, nuts, etc.			to suit
			Total 100.00

Procedure:

PART 1: Heat the **Nulomoline** until dissolved, gradually add the albumen solution and beat until light.

PART 2: Boil the sugar, corn syrup, and water to 255–260 F.

Pour this in a fine steam into the whipped batch (Part 1), and beat until light, then add the fondant and **Nulco MeloBan** (which should be sifted into the batch) or vanilla flavor and color, if desired, Brazil nuts, pecans, etc.

Spread the batch on waxed paper and when it has set, cut into bars, squares, or oblongs or spoon out into kisses, then wrap in transparent or waxed paper or coat with chocolate. If a softer piece is desired, boil Part 2 to 252–255 F.

BASIC FONDANT:		
Granulated sugar	80	lb
Corn syrup	20	lb
Invert sugar	10	lb
Water		

Procedure:

Boil granulated sugar, corn syrup, and invert sugar, plus sufficient water to dissolve the sugar, to 242 F (sea level), then cool and cream into fondant.

Source: Crompton & Knowles

Divinity Kisses

(Friend or Wire Cut Machines)

			% by Wt.
PART 1:			
Nulomoline (invert sugar)	10	lb	10.02
Soy albumen	1	lb	1.00
PART 2:			
Granulated sugar	40	lb	40.25
Corn syrup	20	lb	20.09
Water	10	lb	10.04
Casting fondant	10	lb	10.04
Salt	8	oz	0.50
Sucrovert (liquid invertase)	1	oz	0.06
Candied diced fruits	8	lb	8.00
Color and flavor (white and pink)			to suit
			Total 100.00

Procedure:

PART 1: Heat the **Nulomoline** until dissolved (approximately 150 F), place this into a beater, add the soy albumen and beat until light.

PART 2: Place the sugar, corn syrup, and water into a kettle and stir occasionally until the batch boils. Wash down all crystals of sugar that cling to the kettle and cook the batch to 246 F (at sea level).

Pour this into the whipped portion (Part 1), beating continuously. Add the fondant and salt and continue to beat until the batch becomes short, then add the **Sucrovert**, color, flavor, and mix well. Now add the candied fruits and mix to distribute.

Spread the batch out to form sheets or spoon into the form of kisses, or form on the Friend rolled cream center machine.

For batches to be formed into sheets, cook to 250 F (sea level).

For kisses cook the batch to 246 F.

Source: Crompton & Knowles

Frappé Nougat Kiss

(Chewing)

			% by Wt.
Corn syrup	45	lb	32.73
Nulomoline (invert sugar)	20	lb	14.55
Sugar	35	lb	25.45
Water (to dissolve the sugar)			
Hard coconut butter	5	lb	3.64
Salt	8	oz	0.36
Frappé (q.v.)	32	lb	23.27
			Total 100.00

Procedure:

Cook the corn syrup, **Nulomoline**, sugar, and water to 280 F, turn off the heat, add the frappé and beat well. Add the coconut butter and salt and mix thoroughly; then add flavor and color, as desired. Spread on a cooling table, then cut, size, and wrap or coat with chocolate.

FRAPPÉ:

Corn syrup	50	lb
Nulomoline (invert sugar)	50	lb
Albumen (dissolved in water)	4	lb
Water	8	lb

Procedure:

Boil the corn syrup to 245 F, add the **Nulomoline** (cold) and stir until melted. Place the batch into a marshmallow beater, gradually add the albumen solution and beat the batch until light.

Source: Crompton & Knowles

Grained Nougat-Malo

(To be Coated with Jelly, Icing or Chocolate)

			% by Wt.
PART 1:			
Gelatin (#225 g test)	1.75	lb	1.30
Cold or warm water	11	lb	8.16

PART 2:			
Granulated sugar	65	lb	48.21
Corn syrup	30	lb	22.25
Nulomoline (invert sugar)	10	lb	7.42
Water	15	lb	11.13
PART 3:			
Powdered sugar	2	lb	1.48
Sucrovert (liquid invertase)	1	oz	0.05
Flavor and color			to suit
			Total 100.00

Procedure:

PART 1: Mix well together and set aside.

PART 2: Place the sugar and water into a kettle, stir occasionally until the batch boils, then cook the batch to 236–237 F (sea level). Turn off the heat, add the corn syrup and **Nulomoline** and mix well. Then add the soaked gelatin (Part 1) and mix until the gelatin is dissolved.

Place the batch into a marshmallow beater and beat to obtain a volume of 4 lb of the beaten Nougat-Malo to 1-gal measure.

PART 3: After the batch has been beaten to the required volume, add the **Sucrovert**, powdered sugar, color and flavor, and at this point the Nougat-Malo will register a temperature of approximately 110–115 F.

Now cast the batch into starch impressions at normal room temperature and allow the cast candies to remain in a room at normal temperature until the gelatin has set. Then transfer the starch trays containing the candy into a drying room, which has been tempered to 110–120 F where they should set for approximately 6–10 h or until surface graining has occurred. The candies when sufficiently firm and grained on the exterior may then be removed from the starch and coated with chocolate, icing, or agar coating.

Source: Crompton & Knowles

Light Cast Grained Gelatin Nougat

			% by Wt.
Granulated sugar	65	lb	58.89
Water (to dissolve the sugar)			
Corn syrup	25	lb	22.65

Nulomoline (invert sugar)	10	lb	9.06
Gelatin (medium grade, soaked in 8 lb water)	2.25	lb	2.04
Water	8	lb	7.25
Convertit (invertase)	2	oz	0.11
			Total 100.00

Procedure:

Boil the sugar and water to 236 F. Turn off the heat, add the corn syrup and **Nulomoline,** then add the gelatin which has been soaked in the water for at least $^1/_2$ h and mix until the gelatin is completely dissolved.

Place the batch at once into a marshamallow beater and beat until quite light, then add the **Convertit,** color, and flavor, and cast the nougat into starch which has been tempered to 90–100 F.

Allow the nougat to remain in the starch in a room of normal temperature for several hours, to permit the gelatin to set, then transfer to a drying room tempered to 105–115 F. Allow the nougat to remain in the drying room for approximately 8 h, after which they may be removed from the starch.

Coat the nougat with chocolate, agar, icing or caramel coating, then roll in sugar, coconut, or nutmeats.

Source: Crompton & Knowles

Light Short Nougat

(Cut or Rolled)

			% by Wt.
PART 1—FRAPPÉ (q.v.)	32	lb	27.83
PART 2:			
Sugar	60	lb	52.17
Corn syrup	20	lb	17.39
Water (to dissolve the sugar)			
Hard coconut butter	3	lb	2.61
Chopped preserved fruits, nuts			to suit
Honey flavor			to suit
Salt			to suit
			Total 100.00

Procedure:

Place the frappé (Part 1) in a beater and pour in the hot batch and beat until the mixture grains. Then add hard coconut butter, chopped preserved fruits, nuts, etc. and honey flavor and salt, and mix well.

Turn the batch out to set or pass through a rolled cream center machine.

FRAPPÉ:

Corn syrup	50	lb
Nulomoline (invert sugar)	50	lb
Albumen (dissolved in the water)	4	lb
Water	8	lb

Procedure:

Boil the corn syrup to 245 F, shut off the heat, add the **Nulomoline** and stir until it is melted. Place the batch into a marshmallow beater, start the beater, then gradually add the albumen solution and beat until light.

Source: Crompton & Knowles

Light Soft Nougat Center

			% by Wt.
Whipping agent	2	oz	0.43
Water	2	lb	6.96
Icing	2.5	lb	8.70
Sugar	8	lb	27.82
Water	2.5	lb	8.70
Corn syrup (42 D.E.)	10	lb	34.78
Nonfat dry milk	2	lb	6.96
Icing sugar	10	oz	2.17
Duromel (emulsifier)	1	lb	3.48
			Total 100.00

Procedure:

Mix and beat whipping agent, water, and icing to a stiff foam. Boil sugar, water, and corn syrup to 266 F and add slowly to the stiff foam at low speed. Add nonfat dry milk and icing sugar and mix in. Melt **Duromel** and mix in.

Source: Van den Bergh Foods

Light Chocolate Nougat Center

			% by Wt.
Whipping agent		7 oz	0.77
Water	3.5 lb		6.18
Icing sugar	7 lb		12.36
Sugar	13 lb		22.96
Water	4.5 lb		7.95
Corn syrup (42 D.E.)	20 lb		35.32
Cocoa powder	2 lb		3.53
Nonfat dry milk	4 lb		7.06
Icing sugar	1 lb		1.77
Duromel	1 lb	3 oz	2.10
			Total 100.00

Procedure:

Mix and beat whipping agent, water, and icing sugar to a stiff foam. Dissolve sugar in the water, add corn syrup and boil to 262 F. Add to the stiff foam at low speed. Mix and sift the cocoa powder, dry milk, and icing sugar and add to batch. Melt **Duromel** and mix into batch.

Source: Van den Bergh Foods

MeloBan Sea Foam Blocks and Kisses

		% by Wt.
PART 1:		
Albumen (dissolved in water)	1.25 lb	1.04
Water	3 lb	2.50
Tapioca flour	1 lb	0.83
Nulomoline (invert sugar)	10 lb	8.33
PART 2:		
Granulated sugar	40 lb	33.35
Corn syrup	20 lb	16.67
Water	10 lb	8.33
Basic fondant (q.v.)	10 lb	8.33
Hardened coconut oil (94–96 F m.p.)	3 lb	2.50
Nulco MeloBan (riped banana product mixed with water)	4 lb	3.33

Water	2.5	lb	2.08
Powdered salt	4	oz	0.21
Yellow color			to suit
Raisins	15	lb	12.50
			Total 100.00

Procedure:

PART 1: Place the **Nulomoline** (either warm or cold) into a beater, add the albumen solution and while beating, sift in the tapioca flour. Heat the batch until moderately light.

PART 2: Boil the sugar, corn syrup, and water to 275–280 F (sea level). Gradually add this to Part 1, beating well. Add the fondant and beat until grain is visible, then add the hardened coconut oil, moistened **MeloBan**, salt, and color, mixing well.

Now add the raisins and mix just enough to distribute the raisins, after which the batch may be turned out into waxed paper-lined pans, or it may be formed into sheets or spooned into kisses.

BASIC FONDANT:

Granulated sugar	80	lb
Corn syrup	20	lb
Nulomoline (invert sugar)	10	lb
Water (to dissolve the sugar)		

Procedure:

Boil the sugar, corn syrup, **Nulomoline**, plus sufficient water to dissolve the sugar to 244 F, then cool the syrup to 125 F. Beat the batch into fondant.

Source: Crompton & Knowles

MeloBan Semigrained Nougat

(Cut or Rolled)

			% by Wt.
PART 1: SPECIAL FRAPPÉ (q.v.)	35	lb	24.31

PART 2:

Sugar	60	lb	41.67
Corn syrup	40	lb	27.78
Water (to dissolve the sugar)			to suit

PART 3:

Hard coconut butter	5	lb	3.47
Nulco MeloBan (ripe banana product)	4	lb	2.77
Chopped preserved fruits, nuts, banana color			to suit
Salt			to suit
			Total 100.00

Procedure:

PART 1: Prepare special frappé.

PART 2: Boil altogether to 275 F.

Place into a beater the Special Frappé (Part 1) and while the beater is running add the hot syrup batch (Part 2) and beat until the mixture becomes slightly short.

Add hard coconut butter and **Nulco MeloBan**, chopped preserved fruits, nuts, salt, and banana color and mix well.

Spread the batch out to set and when cold, cut into slices, squares, oblongs, etc. and wrap in transparent cellulose or coat with icing or chocolate.

SPECIAL FRAPPÉ:

Corn syrup	25	lb
Nulomoline (invert sugar)	25	lb
Gelatin (soaked in the water)	12	oz
Water	2	lb
Albumen (dissolved in the water)	2	lb
Water	4	lb

Procedure:

Boil the corn syrup to 246 F, shut off the steam and add the **Nulomoline** and stir until melted. Place the batch into a marshmallow beater. Gradually add the soaked gelatin and beat until the gelatin is fully dissolved. Gradually add the albumen solution and beat the batch until it is light.

Source: Crompton & Knowles

Molasses Nougat

(Cut, Chewing)

			% by Wt.
PART 1—FRAPPÉ:			
Albumen (dissolved in the water)	2	lb	1.72
Water	4	lb	3.45
Nulomoline (invert sugar)	15	lb	12.92
Corn syrup	20	lb	17.23
PART 2—BOB:			
Molasses (**Grandma's**)	50	lb	43.08
Evaporated milk	10	lb	8.62
PART 3:			
Coconut oil	5	lb	4.31
Lecithin	1	oz	0.05
Broken or slice Brazil nuts	10	lb	8.62
			Total 100.00

Procedure:

PART 1—FRAPPÉ: Beat the soaked albumen until very light. Cook the **Nulomoline** and corn syrup to 250 F, then slowly add to the beaten albumen while the beater is in motion.

PART 2—BOB: Cook to 270 F.

Add the bob (Part 2) to the frappé (Part 1) in the beater while the beater is in second speed, then shift to high speed and beat until the desired volume is obtained.

Add coconut oil in which lecithin has been melted. Add broken or sliced Brazil nuts and mix through the batch.

Pour the batch on greased slabs and roll out to the desired thickness and when cool, cut and coat with chocolate.

NOTE:

This nougat should not be beaten any longer than necessary after the fat has been added.

Source: Crompton & Knowles

Nougat Almondine

NOUGAT CENTER:

			% by Wt.
Sugar	4	lb	32.83
Corn syrup	3	lb	24.62
Water	2	lb	16.42
Egg albumen (dissolved in 6 oz water)	3	oz	1.53
Fondant	8	oz	4.10
Melted vegetable fat (92 F)	8	oz	4.10
Roasted chopped almonds	8	oz	4.10
Chopped glacé cherries	8	oz	4.10
Assorted diced fruit	1	lb	8.20
			Total 100.00

Procedure:

Cook sugar, corn syrup, and water to 238 F. Draw off 1 pt and pour in fine stream to egg whip. Beat to a peak (planetary beater). Change wire whip to flat beater and pour in balance of syrup which has been cooked to 272 F on medium speed. After all syrup is in, change to low speed and add fondant, fat, almond, and fruit mixture.

Spread on paper-lined table about $^3/_4$ in. high. When set, cut in bars, dip in caramel and roll in almonds of your choice: roasted or sugar-toasted steel cut.

CARAMEL:

			% by Wt.
Corn syrup	7	lb	32.49
Sugar	5	lb	23.19
Cream (40%)	1	qt	9.83
Evaporated milk	6	lb	27.83
Vegetable fat (92 F)	1	lb	4.64
Vanilla	0.5	oz	0.57
Salt	2	oz	1.45
			Total 100.00

Procedure:

Boil sugar, corn syrup, vegetable fat, salt, and cream while stirring to 235 F. Add evaporated milk a little at a time so as not to stop the boil. Finish cook at 245 F or medium ball.

Source: California Almond Growers Exchange

Nougat Eggs, Bars, Rolls, Etc.

(Cut or Rolled)

		% by Wt.
PART 1:		
Nulomoline (invert sugar)	12.5 lb	10.40
Corn syrup	12.5 lb	10.40
Ground gelatin (soaked in water)	12 oz	0.62
Water	1.5 lb	1.25
Albumen (dissolved in water)	1 lb	0.83
Water	2 lb	1.66
PART 2:		
Granulated sugar	50 lb	41.58
Corn syrup	20 lb	16.63
Water	20 lb	16.63
		Total 100.00

Procedure:

PART 1: Heat the **Nulomoline** to 175 F, add the soaked gelatin and stir until the gelatin is completely dissolved. Add the corn syrup, mix well, and place the batch into an open-type beater. Start the beater, add the albumen solution, and beat the batch until it is very light and stiff.

PART 2: Cook altogether to 256–270 F.

Place the beaten batch (Part 1) into a nougat kettle, then slowly pour the high cooked batch into it, mixing continuously. Continue beating until a slight grain shows, then add color, flavor, nut meats, and cherries and when mixed, spread out at once on a well oiled cooler. When sufficiently cool, roll into eggs, bars, or rolls, and when cold, coat with chocolate or caramel.

Source: Crompton & Knowles

Nougat Montelimart

		% by Wt.
Granulated sugar	3.5 lb	23.48
Regular corn syrup	3 lb	20.18
Honey	2 lb	13.42
Egg albumen (dissolved in 8 oz water)	4 oz	5.00

Whole roasted Mission sheller run almonds	2 lb	13.42
Chopped roasted almonds	8 oz	3.35
Green glacé cherries (chopped)	8 oz	3.35
Red glacé cherries (chopped)	8 oz	3.35
Assorted diced fruit	8 oz	3.35
Diced prunes	8 oz	3.35
Vegetable fat (92 F)	1 lb	6.70
Salt	2 oz	0.84
Vanilla	0.5 oz	0.21
	Total	100.00

Procedure:

Boil honey to 275 F and add in fine stream to whipped egg albumen. Cook sugar, corn syrup, and salt and 2 lb water to 275 F. Add to egg batch in fine stream whipping on medium speed. Reduce to low speed and add fruit and nut mix, vegetable fat, and vanilla. Pour on paper-lined slab or table. When set, cut and wrap.

Source: California Almond Growers Exchange

Nu-Nougatette

(Light Cast Grained Gelatin Nougat)

		% by Wt.
Granulated sugar	65 lb	58.90
Water (to dissolve the sugar)		
Corn syrup	25 lb	22.66
Nulomoline (invert sugar)	10 lb	9.06
Gelatin (medium grade)	2.25 lb	2.03
Water	8 lb	7.24
Sucrovert (liquid invertase)	2 oz	0.11
Color and flavor		to suit
	Total	100.00

Procedure:

Boil the sugar and water to 236 F (sea level). Turn off the heat. Add the corn syrup and **Nulomoline**, then add the gelatin which has been soaked in the 8 lb of water for at least $^1/_2$ h and mix until the gelatin is completely dissolved.

Place the batch at once into a marshmallow beater and beat until light. Then add the **Sucrovert**, color, and flavor; beat well and cast the nougats into starch impressions which have been tempered to 90–100 F. The nougat may be cast into starch impressions on a Thursday or Friday and removed the following Monday, Tuesday, or later, thus utilizing starch trays which otherwise might be idle over the weekend.

The nougats after being cast should remain in a room at normal temperature at least several hours to permit the gelatin to set; then if in a rush, the trays containing the nougat should be transferred to a drying room which has been tempered to 100–115 F, where they should remain for approximately 5–8 h.

The starch trays may then be removed from the drying room and the nougat taken from the starch and coated in chocolate, icing, jelly, or caramel coating.

Source: Crompton & Knowles

Oriental Nougat

(Chewing)

			% by Wt.
Corn syrup	30	lb	41.60
Granulated sugar	15	lb	20.80
Water (to dissolve the sugar)			
Condensed milk	10	lb	13.86
Hard coconut butter	2	lb	2.77
Salt	2	oz	0.17
Frappé (q.v.)	15	lb	20.80
Fruit and flavor			to suit
			Total 100.00

Procedure:

Place the corn syrup, sugar, and waer into a kettle and boil to 280 F (sea level). Now slowly add the condensed milk and coconut butter and cook the batch to 270 F. Turn off the heat and add the frappé and the salt and mix well. Test the batch in cold water and keep adding the frappé and testing the batch until it is of the desired consistency. Add fruit and flavor, as desired.

FRAPPÉ:

Corn syrup	50 lb
Nulomoline (invert sugar)	50 lb
Albumen (dissolved in the water)	4 lb
Cold water	8 lb

Procedure:

Boil the corn syrup to 245 F (sea level), shut off the heat, add the **Nulomoline** and stir until it is melted. Place the batch into a marshmallow beater, start the beater, then gradually add the albumen solution and beat the batch until light.

Source: Crompton & Knowles

Plastic Nougat

(Formed on Friend Rolled Center Machine)

		% by Wt.
PART 1:		
Basic fondant (q.v.)	20 lb	29.00
Frappé No. 2 (q.v.)	15 lb	21.72
Powdered salt	3 lb	4.35
PART 2:		
Granulated sugar	25 lb	36.23
Corn syrup	5 lb	7.25
Water (to dissolve sugar)		
Coconut oil (76 F m.p.)	1 lb	1.45
Vanilla or other flavors		to suit
		Total 100.00

Procedure:

PART 1: Heat the basic fondant to 125 F, add the frappé and salt, and beat in an upright beater until moderately light.

PART 2: Boil the sugar, corn syrup, and water to 270 F (sea level). Pour this into Part 1, beating at second speed until the batch becomes short, then add the coconut oil, vanilla or other flavors and turn the batch out on oiled and dusted trays.

When the batch has cooled, place it into the Friend machine, using either large or small dies, and form into pieces of the desired size and shape. Allow the formed centers to set for several hours or overnight, then coat with Creole Caramel Coating (q.v.) and roll in nuts, toasted coconut, etc.

If genuine banana flavor is desired, add 3 lb of **Nulco MeloBan** and a little yellow color after the batch has been beaten to the desired volume.

BASIC FONDANT:

Granulated sugar	80	lb
Corn syrup	20	lb
Nulomoline (invert sugar)	10	lb
Water (to dissolve the sugar)		

Procedure:

Boil altogether to 240–244 F, cool and beat into fondant in the usual manner.

FRAPPÉ NO. 2:

Corn syrup	50	lb
Nulomoline (invert sugar)	50	lb
Albumen (dissolved in the water)	2	lb
Water (cold)	4	lb

Procedure:

Boil the corn syrup to 245 F. Add the **Nulomoline** at once, mix well, then turn off the heat. Place the batch into a marshmallow beater, start the beater, then gradually add the albumen solution and beat until quite light.

CREOLE CARAMEL COATING:

Condensed whole or skimmed milk	10	lb
Corn syrup	15	lb
Molasses (**Grandma's**)	5	lb
Granulated sugar	5	lb
Powdered salt	1	oz
Coconut oil (76 F m.p.)	1	lb
Basic Fondant (q.v.)	5	lb

Procedure:

Heat the condensed milk, corn syrup, molasses, sugar, salt, and coconut oil until the batch boils, stirring constantly, and continue to stir and cook the batch to 240 F (sea level).

Turn off the heat and immediately add the basic fondant, mixing until the fondant is melted.

Dip the nougat centers in the caramel coating and roll in nut meats, toasted coconut, etc.

Source: Crompton & Knowles

Regular Nougat

		% by Wt.
Egg albumen	0.5 lb	1.58
Water (to cover albumen)		
White sugar	10 lb	31.76
Clear honey	5 lb	15.87
Glucose (corn syrup)	5 lb	15.87
Powdered sugar	5 lb	15.87
Raw almonds, English walnuts, or pignolias	6 lb	19.05
		Total 100.00

Procedure:

Soak $^{1}/_{2}$lb egg albumen in enough water to cover overnight, then stir up well and if all the albumen is dissolved, strain through a sieve and beat up as stiff as possible.

Meanwhile cook white sugar, clear honey, and glucose (corn syrup) to a good ball; take out about 3 lb of this and stir and pour slowly into the egg albumen. This operation should be performed by two persons, one pouring and one stirring well with the egg beater.

Cook balance of the batch to the crack and pour and beat it into the egg batch, then add powdered sugar and beat or stir over hot water bath. Now add raw almonds, English walnuts, or pignolias, fill into wafer-lined trays, and finish. These may be wrapped or coated with chocolate.

In order to make almonds go further, drop them in pieces the size of a large pea, and use 1 lb less then specified.

Source Savage Brothers Co.

Sea Foam Kisses

			% by Wt.
PART 1:			
Albumen (dissolved in water)	1	lb	1.18
Water	3	lb	3.53
Tapioca flour	1	lb	1.18
Nulomoline (invert sugar)	5	lb	5.88
Corn syrup	5	lb	5.88
PART 2:			
Granulated sugar	40	lb	47.06
Nulomoline (invert sugar)	5	lb	5.88
Corn syrup	5	lb	5.88
Water (to dissolve the sugar)			
PART 3:			
Basic fondant	20	lb	23.53
Salt, flavor, color, fruits, and nuts			to suit
			Total 100.00

Procedure:

PART 1: Place the dissolved albumen, tapioca flour, **Nulomoline**, and corn syrup into a beater and beat until light.

PART 2: Boil the sugar, **Nulomoline**, corn syrup, and water to 248 F. Gradually add about 15 lb of this syrup to Part 1, beating continuously. Allow the batch to beat for a minute or two, then gradually add the balance of the cooked syrup and beat until light.

Add the basic fondant, salt, color, flavor, fruits and nuts, and mix thoroughly.

Spoon or deposit the candy onto rubber mats, waxed paper, etc. Allow the candies to set until they dry on the surface, after which they are ready to be packed.

Source: Crompton & Knowles

Sea Foam Sheets

			% by Wt.
PART 1:			
Albumen (dissolved in water)	1.25	lb	1.50
Water	3	lb	3.60
Tapioca flour	1	lb	1.20
Nulomoline (invert sugar)	10	lb	12.01
PART 2:			
Granulated sugar	35	lb	42.06
Water (to dissolve the sugar)			
Corn syrup	20	lb	24.02
PART 3:			
Basic fondant (q.v.)	10	lb	12.01
Salt			to suit
Flavor, color, nuts			to suit
Hardened coconut oil	3	lb	3.60
			Total 100.00

Procedure:

PART 1: Place the **Nulomoline** into a beater, add the dissolved albumen and beat until moderately light. Sift in the tapioca flour and beat well.

PART 2: Boil the sugar, water, and corn syrup to 270–275 F (sea level) and gradually add this to Part 1, beating until moderately light.

PART 3: Break the basic fondant apart and drop about 1 lb pieces into the beaten batch. Add the flavor, salt, color, etc. and beat until the batch has grained, then add the hardened coconut oil which has been cut into small pieces and mix just enough to thoroughly distribute and melt the coconut oil.

Spread the batch on dusted oiled paper or waxed paper. Nuts, pectin jellies or dried fruits may also be added.

BASIC FONDANT:		
Granulated sugar	80	lb
Corn syrup	20	lb
Nulomoline	10	lb
Water (to dissolve the sugar)		

Procedure:

Cook the sugar, corn syrup, **Nulomoline,** and sufficient water to 240–244 F, then cool and beat into fondant.

Source: Crompton & Knowles

Sea Foam Sheets or Kisses

			% by Wt.
PART 1:			
Egg albumen	1	lb	1.28
Water	2	lb	2.56
Nulomoline (invert sugar)	10	lb	12.81
PART 2:			
Granulated sugar	25	lb	32.03
Water	10	lb	12.81
Corn syrup	20	lb	25.62
Casting fondant	10	lb	12.81
Convertit (invertase)	1	oz	0.08
Flavor and color			to suit
Nut meats, diced preserved fruits			to suit
			Total 100.00

Procedure:

PART 1: Dissolve the albumen in the water and place this into a beater. Add the **Nulomoline** (cold) and beat the batch until quite light.

PART 2: Place the sugar, water, and corn syrup into a kettle and mix well. Apply heat and occasionally stir the batch until it boils and cook to 245–250 F (sea level), then pour this in a fine stream into the whipped portion (Part 1), beating at second speed.

After all of the cooked syrup has been added, beat for a minute or two, then add the fondant, beating until the batch becomes plastic. Now add the **Convertit,** flavor, and color, mixing well, then add the nut meats, preserved fruit, etc., distributing well through the batch.

If the batch is to be formed into sheets or layers, cook to approximately 255 F.

For kisses, cook the batch to approximately 250 F.

Source: Crompton & Knowles

Semichewing Nougat

No. 1

(Cast into Starch)

		% by Wt.
PART 1—SPECIAL FRAPPÉ:		
Corn syrup	30 lb	18.24
Nulomoline (invert sugar)	20 lb	12.20
Albumen (dissolved in the water)	1.5 lb	0.91
Water	3 lb	1.82
PART 2—NOUGAT BATCH:		
Granulated sugar	45 lb	27.31
Corn syrup	35 lb	21.28
Water	10 lb	6.08
Basic Fondant (q.v.)	20 lb	12.16
		Total 100.00

Procedure:

PART 1: Boil the corn syrup to 255 F. Turn off the heat, add the **Nulomoline**, and stir until the **Nulomoline** is melted. Place the batch into a marshmallow beater, start the beater, then gradually add the albumen solution and beat the batch until it is quite light.

PART 2: Boil the sugar, corn syrup, and water to 260 F. Pour this gradually into the entire amount of Special Frappé (Part 1). Mix thoroughly while pouring, then add the basic fondant which has been melted before being mixed with the batch. Add 4 oz of salt and flavor and color as desired.

Meanwhile have the starch trays printed in bar forms. Drop nuts or fruit into the starch impressions, then cast the nougat on top. Allow to set until the following morning, then handle in the usual manner. (For each 550 ft above sea level reduce the boiling temperature by 1 F.)

BASIC FONDANT:	
Sugar	80 lb
Corn syrup	20 lb
Water (to dissolve sugar)	

Procedure:

Cook altogether to 242 F, then cool and beat into fondant.

Source: Crompton & Knowles

No. 2

(For Cutting or Casting)

			% by Wt.
PART 1—FRAPPÉ NO. $2^1/_2$	35	lb	21.11
PART 2:			
Granulated sugar	50	lb	30.17
Corn syrup	50	lb	30.17
Water (to dissolve the sugar)			
PART 3:			
Basic fondant (q.v.)	20	lb	12.07
Prep-Co (plastic coconut)	10	lb	6.03
Powdered salt	12	oz	0.45
Flavor, color, nuts			to suit
			Total 100.00

Procedure:

PART 1: Prepare frappé. Place the frappé into a nougat beating kettle.

PART 2: Boil the sugar, corn syrup, and water to 260–265 F (sea level), then gradually pour this into the frappé, beating at second speed, then add Part 3.

PART 3: Mix the batch thoroughly until the fondant is melted, adding color, flavor, nuts, etc. as desired, and mix well.

Cast the batch into starch impressions or spread on an oiled slab, rolling the batch to the required height and when it has set, cut into pieces of the desired size and wrap in waxed paper, cellophane, or coat with chocolate.

FRAPPÉ NO. $2^1/_2$:		
Corn syrup	25	lb
Nulomoline	25	lb
Albumen (dissolved in the water)	1.25	lb
Water	2.5	lb

Procedure:

Boil the corn syrup to 245 F. Add the **Nulomoline** and mix unti it is melted.

Promptly transfer this syrup to a marshmallow beater, start the beater, then gradually add the albumen solution and beat the batch until light.

BASIC FONDANT:		
Granulated sugar	80	lb
Corn syrup	20	lb
Nulomoline	10	lb
Water (to dissolve the sugar)		

Procedure:

Boil the sugar, corn syrup, and **Nulomoline** with sufficient water to dissolve the sugar, heat to 244 F, cool the syrup to 125 F, then beat into fondant.

Source: Crompton & Knowles

Short Chewing Nougat

			% by Wt.
PART 1:	25	lb	21.74
Nulomoline (invert sugar)	25	lb	
Corn syrup	25	lb	
Ground gelatin (wet with 3 lb cold water)	1.5	lb	
Albumen (dissolved in 4 lb water)	2	lb	
PART 2:			
Granulated sugar	50	lb	43.48
Corn syrup	20	lb	17.39
Water	20	lb	17.39
			Total 100.00

Procedure:

PART 1: Heat the **Nulomoline** to 175 F, add the soaked gelatin, and stir until the gelatin is completely dissolved; then add the corn syrup, mix well, and place the batch into an open-type marshmallow beater. Start the beater, and add the albumen which has been previously dissolved. Beat the batch until it is very light and stiff.

PART 2: Cook altogether to 265 F for cold weather and in hot weather increase the cooking temperature to 270 F.

Have in the nougat kettle 25 lb of Part 1 and pour the high cooked batch slowly into it mixing continuously. Continue beating until a slight grain shows, then add color, flavor, and nuts, and when mixed, spread out at once on a well oiled cooler.

Roll the batch at once using bars that will give the thickness wanted. To avoid sticking have the rolling pin oiled and then dusted with starch. It is advisable to cut this batch while warm. If for any reason this cannot be done, then 4 lb of hard coconut butter should be added to the batch while the high cooked batch is being poured into Part 1.

This makes an excellent wrapped piece and it can also be coated with chocolate or iced.

Source: Crompton & Knowles

Summer Nougat

		% by Wt.
PART 1:		
Albumen (dissolved in the water)	2.5 lb	1.30
Water	6 lb	3.15
Tapioca flour	2 lb	1.05
Nulomoline (invert sugar)	20 lb	10.50
PART 2:		
Granulated sugar	70 lb	36.75
Water	20 lb	10.50
Corn syrup	40 lb	21.00
PART 3:		
Basic fondant (q.v.)	20 lb	10.50
Salt		to suit
Hardened coconut oil	10 lb	5.25
Flavor, color, nuts		to suit
		Total 100.00

Procedure:

PART 1: Place the **Nulomoline** into a beater, add the dissolved albumen, and beat until moderately light. Sift in the tapioca flour and beat well.

PART 2: Cook the sugar, water, and corn syrup to 270–275 F (sea level) and gradually add this to Part 1, beating until moderately light.

PART 3: Break the basic fondant apart and drop about 1-lb pieces into the beaten batch. Add the salt, flavor, color, etc., and beat until the batch has grained, then add the hardened coconut oil which has been cut into small pieces, and mix just enough to thoroughly distribute and melt the coconut oil.

Spread the batch on dusted, oiled or waxed paper. Nuts, pectin jellies, or dried fruits may also be added.

Source: Crompton & Knowles

Chapter 12

NOTABLE AND NOVEL CONFECTIONS

The formulas that follow represent a random selection of candies and confections that fall in no special category but are unique in their own character—flavor, color, texture, and shape.

Almond Popcorn Crunch

			% by Wt.
White sugar	5	lb	25.99
Brown sugar #10	2	lb	10.39
Regular corn syrup	2.5	lb	12.99
Water	3	lb	15.58
Salt	4	oz	1.29
Roasted Mission sheller run almonds	4.5	lb	23.38
Butter	12	oz	3.89
Popped corn	1.25	lb	6.49
			Total 100.00

Procedure:

Cook sugar, corn syrup, salt, and water to 230 F. Add almonds and cook slowly to 255 F. Continue to cook but begin stirring. At 295 F add butter and cook until butter is incorporated. Shut off fire and stir in popped corn. Spread on slab and separate with oiled, gloved hands.

Source: California Almond Growers Exchange

Bark

			% by Wt.
Butterscotch coating	3	lb	82.76
Bitsyn® or nut pieces	10	oz	17.24
			Total 100.00

Procedure:

Temper coating by heating to 125 F, cooling to 88 F, and reheating to 95 F. Add **Bitsyn®**, mix well, and pour onto waxed paper. Spread thin ($^1/_8$ to $^1/_4$ in.), roughen top, and allow to dry. Cut in pieces approximately $5^1/_2 \times 8$ in. just before setting.

Source: Pillsbury Co.

Candy Cigarette Dough

			% by Wt.
Powdered sugar	81.57	lb	71.12
Clintose brand dextrose (VF granulation)	24.47	lb	21.34
Clinton regular corn syrup (42 D.E.)	8.16	lb	7.12
Gum tragacanth	0.24	lb	0.21
Gelatin	0.24	lb	0.21
			Total 100.00

Procedure:

Gum tragacanth is made up on the basis of 4 lb of gum to 5 gal of water. One pound of gelatin is dissolved in 1 gal of water. Corn syrup, gum, **Clintose**, and sugar are added to the mixer in the order named. Start the mixer and then add gelatin solution. Mix until dough reaches desired consistency. Flavor should be added just before the dough reaches the desired consistency.

NOTES:

Formula quantities of water may have to be varied somewhat depending on the equipment used.

Source: ADM

Caramel Corn/Bitsyn® Bars

(Fiddle-Faddle Type)

			% by Wt.
Brown sugar	3	oz	6.14
White sugar	17	oz	34.83
Regular corn syrup	8.5	oz	17.41
Water	8	oz	16.38
Salt	7	g	0.51
Salted dairy butter	2	oz	4.09
Lecithin	2.5	g	0.18
Popped popcorn*	5	oz	10.23
Bitsyn® PC 1765*	5	oz	10.23
			Total 100.00

Procedure:

Cook together in medium-sized copper ketle the sugar, corn syrup, water, and salt to 290 F. Add butter and lecithin, cook back to 310 F. Remove from fire, add popcorn and **Bitsyn®** and mix to coat well. Pour onto warmed, sprayed slab and turn over a few times. Press down with sheet and cut with knife to approximately $1 \times 4^1/_2 \times {}^3/_4$ in. bars.

Source: Pillsbury Co.

Jumbo Cherries

(Flexible Cast Marshmallow, Coated in Agar Jelly, Rolled in XXX Sugar, Coconut, etc.)

			% by Wt.
Nulomoline (invert sugar)	50	lb	46.73
Corn syrup	50	lb	46.73
Gelatin (200 bloom)	2	lb	1.87
Water	5	lb	4.67
			Total 100.00

* Mix and heat

Procedure:

Soak the gelatin in the water. Heat the **Nulomoline** and corn syrup to approximately 200 F, and then add the soaked gelatin and stir until it is dissolved. Place the batch into a marshmallow beater and beat until quite light, then cast.

If the weather or workroom is damp, it is advisable to sift some starch over the tops of the trays. Leave the goods in the starch until the following day, then remove them and coat with chocolate, jelly, or icing.

Source: Crompton & Knowles

Maple Cream

			% by Wt.
Maple syrup	85	lb	85.00
Maple invert sugar	15	lb	15.00
			Total 100.00

Procedure:

Boil together to 238 F. Cool the syrup to approximately 120 F, and then beat it into fondant, and pack into glass or metal containers.

Source: Crompton & Knowles

Maple Sugar Cakes

			% by Wt.
Genuine maple sugar syrup	90	lb	90.00
Maple sugar invert syrup	10	lb	10.00
			Total 100.00

Procedure:

Boil together to 250 F, then cool this syrup to 200 F. Add 5 lb of the maple cream (q.v.), mix well, and cast into rubber molds, cupcake pans, or into paper cups.

Source: Crompton & Knowles

Sweet/Sour Tablets

			% by Wt.
Royal-T dextrose with maltodextrin	97.00	lb	97.00
Malic or citric acid	1.82	lb	1.82
Magnesium stearate	1.00	lb	1.00
Flavoring	0.12	lb	0.12
Color	0.06	lb	0.06
			Total 100.00

Procedure:

Add color to dextrose and blend until uniform. Add remaining ingredients and blend. Press into tablets.

Source: CPC International

Marzipan Candies

			% by Wt.
Almond paste	15	lb	37.38
Short or standard casting fondant	10	lb	24.92
Finely powdered sugar	15	lb	37.38
Invertase	1	oz	0.16
Salt, color, and flavor	1	oz	0.16
			Total 100.00

Procedure:

Mix above ingredients to obtain a paste sufficiently firm to retain its shape when the marzipan is formed. The basic almond paste, fondant, sugar, and invertase may be mixed to form a masterbatch; color and flavor may be added later.

BONBON CENTERS: The marzipan may be rolled into sheets $^3/_8$ to $^1/_2$ in. high, then cut into shapes. Powdered sugar may be used when forming. Place the cut shapes on trays lined with waxed paper and permit them to dry on the surface before dipping in fondant cream or chocolate.

CRYSTALLIZED MARZIPAN: Marzipan may be colored and flavored with vanilla, orange, lemon, Roman Punch, pistachio, or rum flavors. Dust with powdered sugar when rolling or cutting. The pieces are then permitted to dry on the surface before dipping in cool crystallizing syrup, cooking to approximately $33^1/_2$ Bé (at boiling point). The crystallizing syrup is made by heating 72 lb of sugar with 28 lb of water to the boiling point. Test while boiling to obtain a Baumé of $33^1/_2$.

The surface of the crystallizing syrup is sprinkled lightly with cool water and the syrup is allowed to cool to 85–75 F, then ladled over the marzipan candies which have been placed in wire baskets of crystallizing pans. Metal screens are placed over the candies to keep them immersed in the syrup. The candies are left in the syrup for 7–12 h, after which the syrup is drained and the candies turned out on heavy waxed paper to dry.

Source: California Almond Growers Exchange

Marzipan Centers

(Coat with Bonbon Cream or Chocolate or Crystallize)

			% by Wt.
Almond paste	15	lb	33.29
Short bonbon fondant	15	lb	33.29
Finely powdered sugar	15	lb	33.29
Sucrovert (invertase)	1	oz	0.13
Violet, vanilla, orange, rum, or pistachio flavor			to suit
Color			to suit
			Total 100.00

Procedure:

Knead or mix all of the above ingredients together to form a firm mass or dough. The batch may then be divided into various portions and each portion flavored and colored differently.

If desired, the marzipan may be rolled into a sheet, then cut with a standard hand lozenge cutter, or the sheet may be cut into squares, oblongs, and coated with fondant or chocolate.

Two or more layers of various colors may be rolled or sized, then cut into squares or oblongs. If desired, the marzipan mass may be rolled into pieces and then formed into various shapes with the aid of standard marzipan molds or rubber molds as used for candy cream or conserve centers.

The marzipan may also be crystallized in a cool crystallizing syrup of $33^1/_2$ Bé.

Source: Crompton & Knowles

Turkish-Style Delight

(For Cutting)

			% by Wt.
Granulated sugar	60	lb	26.13
Nulomoline (invert sugar)	40	lb	17.42
Thick boiling corn starch	9.5	lb	4.14
Water	120	lb	52.26
Cream of tartar	2	oz	0.05
Orange, mint, raspberry, lemon flavor, etc.			to suit
			Total 100.00

Procedure:

Place the granulated sugar, **Nulomoline,** and half the water into a kettle and heat the batch until it boils.

Meanwhile, mix the corn starch with the remainder of the water (cold) and gradually feed the starch–water mixter to the batch, keeping the batch constantly boiling. Add the cream of tartar and continue to cook and stir the batch to a rather firm jelly string.

Turn off the heat, add the flavor and coloring, mix thoroughly, then pour the batch into a tray which has first been lined with heavy wrapping paper.

Sprinkle the surface of the batch with dry corn starch and allow it to set for 2–3 days after which the paper may be removed by moistening it with water, then sprinkle the batch well with a combination of 5 lb of powdered, dry corn starch, mixed with 1 lb of dry powdered sugar.

Cut the jelly into squares or oblongs, rolling it in the sugar–starch combination. Allow the candies to set for several hours or overnight, then pack into containers.

Source: Crompton & Knowles

Chapter 13

PAN-COATED CANDIES

Pan-Coated Cordial Cherries

		% by Wt.
Drained maraschino-type cherries		
(1100–1200 count)	6 lb	32.16
Nulofond	12 lb	64.33
Drained cherry syrup	10 oz	3.35
Sucrovert (liquid invertase)	0.5 oz	0.16
		Total 100.00

Procedure:

Drain the cherries for 5–10 min. Then place the drained cherries into the revolving pan, add approximately $1^1/_2$ lb of **Nulofond**, start the pan in motion, permit the pan to revolve until the cherries have taken up all of the **Nulofond** and are rolling free. Use more or less fondant as required to lightly coat the cherries.

After the cherries have been coated, mix the **Sucrovert** with the drained cherries with the syrup mixture. Apply just enough of the syrup to lightly moisten the cherries, and when they begin to cluster or cling together, add an additional $^1/_2$ lb (more or less) of the **Nulofond.** When the cherries are rolling freely, apply another charge of the syrup, just sufficient to lightly moisten the cherries, after which an additional $^1/_2$ lb of **Dri-Fond** is added. Repeat this operation until all of the syrup and **Nulofond** have been applied.

If desired, rum, Roman Punch, or other flavors may be added to the syrup before it is sprayed over the cherries. As soon as the cherries are properly coated with the syrup and **Nulofond**, place them into pans or trays in a single layer. The cherries are now ready for final coating.

This formula is based upon the addition of 2 lb of **Dri-Fond** for each 1 lb of drained cherries. If desired, either larger or smaller percentages of **Dri-Fond** may be applied to the cherries.

NOTE:

Speed of small 18 in. revolving pan 20–24 rpm.

Source: Crompton & Knowles

Chapter 14

SYRUPS AND TOPPINGS

Syrups and toppings play an important role in the production of confections and in recent years have enjoyed a remarkable growth. An important function of syrup is its use in controlling the granulation of a batch of candy. Toppings are used to enhance the flavor and appearance of confections.

Low Crystallizable Syrups

Concentration %	Baumé cold at 68 F	Sugar lb	**Nulomoline** lb	Water lb
65	35	53	16	31
66	35.5	51	20	29
67	36	49.5	23.5	27
68	36.5	48	27	25
69	37	46.5	30	23.5
70	37.5	45	33.3	21.7
71	38	43.5	36.7	19.8
72	38.5	42	40	18
73	39	40.5	43.3	16.2
74	39.5	39	46.7	14.3
75	40	37.5	50.0	12.5
76	40.5	36	53.3	10.7
77	41	34.5	56.7	8.8
78	41.5	33	60.0	7
79	42	31.5	63.3	5.2
80	42.5	30	66.7	3.3

The previous table gives the number of pounds of sugar, **Nulomoline**, and water required to produce one hundred pounds of low crystallizable syrups of different concentrations ranging from 65 to 80%.

Procedure:

Heat must be applied to the given mixture to dissolve the sugar. Only moderate heat should be used and the mixture must not be boiled as this would evaporate part of the water and the syrup might crystallize due to low water content. After the syrup has been prepared and cooled to 68 F it should be tested for Baumé as a check against any possible error in weighing the material.

Source: Crompton & Knowles

Bittersweet Topping

			% by Wt.
Water	29	lb	16.78
Granulated sugar	41	lb	23.73
Nulomoline (invert sugar)	78	lb	45.15
Cocoa powder	17	lb	9.84
Unsweetened chocolate	7	lb	4.05
Fine table salt	12	oz	0.43
Vanillin crystals			0.02
			Total 100.00

Procedure:

Heat the sugar and water to 175 F. At this point the sugar should be completely dissolved, then add the **Nulomoline** and stir until the **Nulomoline** is dissolved. Sift in the cocoa powder and add the chocolate, which should be melted or broken into small pieces. Mix thoroughly until the chocolate is dissolved, then add the salt and vanillin crystals and continue to mix and heat the batch until it registers a temperature of 190 F. Turn off the heat and pack the syrup at once into suitable containers.

Source: Crompton & Knowles

Butterscotch Syrup

			% by Wt.
Butter	5	lb	16.67
Sugar	5	lb	16.67
Nulomoline (invert sugar)	20	lb	66.66
Salt			to suit
			Total 100.00

Procedure:

Dissolve sugar in water, bring to a boil, add butter, and cook to 300 F. Then add **Nulomoline** and salt.

Source: Crompton & Knowles

Butterscotch Topping

			% by Wt.
Corn syrup	10	lb	33.06
Nulomoline (invert sugar)	10	lb	33.06
Salted dairy butter	3	lb	9.92
Sweetened condensed whole milk	5	lb	16.53
Molasses (**Grandma's**)	2	lb	6.61
Corn starch	2	oz	0.41
Salt	2	oz	0.41
			Total 100.00

Procedure:

Place the corn syrup, **Nulomoline**, butter, condensed milk into a kettle, then stir and cook altogether to 260 F.

Allow the batch to remain on the fire with the heat off for approximately 5 min in order to permit the batch to develop additional color. Then add the molasses, water, starch, and salt; mix well, and, if necessary, add additional water to reduce the batch to the proper consistency.

Source: Crompton & Knowles

Caramel Paste for Sundae Topping

			% by Wt.
Condensed whole milk (40% sugar)	75	lb	35.63
Fine granulated sugar	25	lb	11.88
Nulomoline (invert sugar)	40	lb	19.00
Corn syrup	50	lb	23.75
Salted dairy butter	10	lb	4.75
Coconut butter (86 F m.p.)	10	lb	4.75
Salt	8	oz	0.24
Vanilla and bergamot flavors			to suit
			Total 100.00

Procedure:

Place the corn syrup, **Nulomoline**, coconut butter, dairy butter, and condensed milk into a kettle, turn the heat on partly and mix continuously until the fats are completely dissolved, then add the sugar and salt.

Heat and stir the batch until the sugar is dissolved but do not permit the batch to boil. When the sugar is completely dissolved, turn off the heat, add the flavors, mix well, and when the batch has cooled to approximately 120 F it is ready to be packed into containers.

This preparation may be used as a caramel topping for sundaes by diluting it with simple syrup, milk, or cream.

Source: Crompton & Knowles

Coconut Nut Topping

			% by Wt.
Sugar	15	lb	52.98
Nulomoline (invert sugar)	1	lb	3.53
Water (to dissolve the sugar)			
Desiccated coconut	8	lb	28.26
Ground Brazil nuts, etc.	4	lb	14.13
Salt	1	oz	0.22
Butter	4	oz	0.88
Red color			to suit
			Total 100.00

Procedure:

Boil the sugar, **Nulomoline**, and water to 285 F, add the butter and color, mix well, then add the coconut, nut meats, etc. and salt and stir until the batch grains. Now pass the batch through sieves of various sizes in order to secure uniform particles of two or three different size toppings.

Source: Crompton & Knowles

Chocolate Milk Syrup

			% by Wt.
Water	20	lb	30.52
Sodium alginate	3.25	oz	0.31
Cocoa powder (natural process 18% fat)	2.25	lb	3.43
Granulated sugar	20	lb	30.53
Powdered milk	1	lb	1.53
Nulomoline (invert sugar)	22	lb	33.58
Powdered salt			to suit
Benzoate of soda or potassium sorbate (powdered)	1	oz	0.10
Vanilla flavor			to suit
			Total 100.00

Procedure:

Heat all of the water to 190–195 F. Sift in the sodium alginate and mix until the alginate is dissolved.

Mix the cocoa powder with the granulated sugar and add this to the algiante solution, mixing until smooth. Sift in the powdered milk, mixing until smooth. Add the **Nulomoline,** salt, and benzoate of soda, which has been dissolved in a little water, and heat the batch gradually to 212–215 F.

Turn off the heat, add the vanilla flavor, mixing thoroughly. The batch may also be allowed to cool and the vanilla flavor added and mixed well.

This syrup has a calculated solids content of 62% before processing and a finished solids content of 68%.

This is an excellent chocolate milk syrup, fairly strong in milk flavor and of moderate cocoa flavor. It will remain stable in body for a period of approximately 1 month whether homogenized or not.

Source: Crompton & Knowles

Chocolate Syrup

No. 1

		% by Wt.
Cocoa powder (natural process 18% fat)	4 lb	13.32
Water	7 lb	23.31
Nulomoline (invert sugar)	13 lb	43.29
Granulated sugar	6 lb	19.98
Powdered salt		to suit
Benzoate of soda (powdered) or potassium sorbate	0.5 oz	0.10
Vanilla flavor		to suit
		Total 100.00

Procedure:

Mix the cocoa powder with the water and allow to soak for approximately 15 min.

Add the **Nulomoline** and heat the batch to 185 F, then add the granulated sugar, salt, and benzoate of soda, which has been dissolved in a little water, and gradually heat the batch to 215 F, holding at this temperature for approximately from 3–5 min.

If it is desired to pack the syrup hot, add the vanilla flavor, mixing well, then pack.

If it is not necessary to pack the syrup hot, allow it to cool, then add the vanilla flavor, mixing thoroughly, and pack.

This chocolate syrup has a calculated solids content of approximately 66% when completed. It has excellent character and flavor and will remain stable for at least a month, either homogenized or not. If the syrup is homogenized, it will have a smoother body and will be of slightly lower viscosity.

Source: Crompton & Knowles

No. 2

(Cocoa Powder)

		% by Wt.
Cocoa powder	12 lb	6.98
Water (cold)	70 lb	40.70

Granulated sugar	35	lb	20.35
Nulomoline (invert sugar)	55	lb	31.97
Flavor			to suit
			Total 100.00

Procedure:

Mix the cocoa powder with the cold water until it is free of lumps, then boil the batch until it is fairly thick. Add the sugar and **Nulomoline** and boil the batch to the proper consistency (72° Brix). Add flavor, mix well, and pack.

Source: Crompton & Knowles

No. 3

			% by Wt.
Cocoa powder (natural process 18% fat)	10	lb	16.93
Water	15	lb	25.40
Nulomoline (invert sugar)	22	lb	37.25
Granulated sugar	12	lb	20.32
Powdered salt			to suit
Benzoate of soda (powdered) or potassium sorbate	1	oz	0.10
			Total 100.00

Procedure:

Mix the cocoa powder and water together and allow to soak for approximately 15 min.

Add the **Nulomoline** and heat the batch to 135 F. Add the sugar, salt, and benzoate of soda, which has been dissolved in a little water, and heat the batch to 212–215 F (sea level) and hold at this temperature for 3–5 min.

If it is desired to pack the syrup hot, add the vanilla flavor, mix well, and pack. If it is not necessary to pack the syrup hot, allow it to cool, then add the vanilla flavor, mixing thoroughly, and pack.

The calculated solids content of the mixed syrup is approximately 63%. This formula makes a syrup of strong flavor and remains fairly stable for a month whether homogenized or not. However, if the syrup is homogenized, it will be smoother and less viscous.

Source: Crompton & Knowles

Commercial Butterscotch Topping

			% by Wt.
Corn syrup	70	lb	46.88
Nulomoline (invert sugar)	30	lb	20.09
Thin boiling starch	3	lb	2.01
Water	25	lb	16.74
Salted dairy butter	6	lb	4.02
Sweetened condensed milk (skim or whole)	15	lb	10.05
Salt	5	oz	0.21
			Total 100.00

Procedure:

Cook the water (cold), starch, and **Nulomoline** to approximately 118 F (jelly string), add the corn syrup, butter, and condensed milk, and cook to a moderate consistency (about 3 min), add the salt and additional flavoring, as desired, then cool and pack.

Source: Crompton & Knowles

Crushed Butter Toffee for Ice Cream

			% by Wt.
Sugar	10	lb	44.33
Nulomoline (invert sugar)	1.5	lb	6.65
Water	1	pt	4.43
Dairy butter	2	lb	8.86
Coconut oil (76 F m.p.)	6	lb	26.59
Caramel paste	2	lb	8.86
Salt	1	oz	0.28
			Total 100.00

Procedure:

Place the sugar, **Nulomoline**, water, dairy butter, and coconut oil into a kettle, mix well and when the batch begins to boil, carefully wash down all grains of sugar that adhere to the interior of the kettle. Boil the batch to 280 F, then add the caramel paste and continue to cook the batch to 300–310 F; add the salt, mix well and spread the batch out and when cold, break it into small pieces.

If pecans are to be used, simply stir them into the batch after it has been cooked to the desired temperature. If almonds are to be used in the toffee, it would be advisable to add the almonds at approximately 280 F, so that they will become slightly roasted. If pecans are to be used, simply stir them into the batch after it has been cooked to the desired temperature.

Source: Crompton & Knowles

Fudge Topping

			% by Wt.
Unsweetened chocolate	5	lb	11.11
Cocoa powder	5	lb	11.11
Water	15	lb	33.33
Sugar	10	lb	22.22
Nulomoline (invert sugar)	10	lb	22.23
			Total 100.00

Procedure:

Mix the cocoa powder with the water then add the unsweetened chocolate which should be melted. Immediately add the sugar and **Nulomoline**, mixing well. Before adding the sugar and **Nulomoline**, the batch will be quite stiff, but will thin out considerably upon addition of these ingredients.

Heat the batch until it boils, stirring constantly and continue to boil until it falls off the spatula in pasty sheets.

Allow the batch to cool, then add vanilla flavor and, if necessary, thin it with water.

Source: Crompton & Knowles

Genuine Maple Invert Syrup

			% by Wt.
Genuine maple sugar syrup	100	lb	99.00
Sucrovert (liquid invertase, 450 cc)	1	lb	1.00
			Total 100.00

Procedure:

Mix the 1 lb of **Sucrovert** with a gallon of the maple syrup, then add this to the remainder of the maple syrup, mixing thoroughly.

Hold this at temperatures varying between 65 and 75 F for at least 15 days. The result should be a genuine maple invert syrup of 90–95% inversion.

Maple sugar candy may be made using a portion of the maple invert.

Source: Crompton & Knowles

Heavy Marshmallow Topping

(5 lb/gal)

		% by Wt.
Corn syrup	75 lb	74.26
Nulomoline (invert sugar)	25 lb	24.75
Albumen (dissolved in the water)	1 lb	0.99
Water	3 lb	
		Total 100.00

Procedure:

Heat the corn syrup to approximately 175 F, add the **Nulomoline** and mix well. Place the batch at once into a marshmallow beater, start the beater, add the properly dissolved albumen and beat to secure a volume equal to 5 lb to the American gallon, then pack into suitable containers.

Source: Crompton & Knowles

Marshmallow Topping

		% by Wt.
Corn syrup	50 lb	44.44
Fine granulated sugar	25 lb	22.22
Water	10 lb	8.88
Nulomoline (invert sugar)	25 lb	22.22
Gum karaya	3 oz	0.18
Albumen (dissolved in 5 lb water)	2.5 lb	2.00
Citric acid (20% sol'n.)	1 oz	0.06
Flavor		to suit
		Total 100.00

Procedure:

Place sugar, corn syrup, and 10 lb of water into a kettle and cook altogether to 236 F (sea level), then shut off the steam. In the meantime, mix the gum karaya with the **Nulomoline** until it is smooth, then add this mixture to the boiled batch and stir until melted. Place the batch into a marshmallow beater, start the beater, then gradually add the albumen solution. While beating add the acid solution and flavor and beat until one U.S. gallon measure when filled with the marshmallow weighs 4 lb.

Source: Crompton & Knowles

Chapter 15

GLOSSARY

Absolute temperature. Temperature given on the Kelvin scale, which is equal to Celsius temperature plus 273; 0°C = 273°K.

Acid. A substance that furnishes hydrogen ions in solution. Typical acids used in candy are citric, malic, and tartaric acids.

Aerated confections. Supersaturated solutions that form grained confections such as creams, fondants, nougats, fudges, and grained marshmallow items.

Agar. Gelatinous carbohydrate sometimes used for thickening jelly candies.

Alkali. A strong water-soluble base.

Amino acids. Essential structural units of protein.

Amorphous. Lacking crystalline structure.

Amylopectin. Branched glucose polysaccharide structure, found particularly in waxy maize starch (N = 24 to 30 glucose units).

Antioxidants. Ingredients used to preserve freshness or prevent oxidative rancidity in candies. Examples are BHA, BHT, TBHQ, and propyl gallate.

A.O.M. (Active Oxygen Method). Air is bubbled through a sample of 500–1000 g of fat under standard conditions until a level of 100 meq of peroxide is produced. Reported in hours of stability.

Aw. Water activity; defined as water vapor pressure of H_2O in product to vapor pressure of pure H_2O at same temperature (25 C). It has values from 0.00–1.00. 0 = dry environment of anhydrous lactose equilibrated with anhydrous P_2O_5. 1 = pure water. Microbes are inhibited below Aw 0.60, as in toffee candy.

Balling. Form, when sieving a powder, agglomerates (balls) develop that cannot pass through the mesh of the sieve.

Base. An oxide or hydroxide of a metal, e.g., NaOH.

°Baumé (Bé). (Modulus 145). Measures percentage of solids. 61° Brix = 33.0 ° Bé.

Baumé scale. A scale devised for convenient hydrometer reading that can be related to concentrations of sugar solutions.

Beat. To stir vigorously by hand or by other mechanical means.

Blanching. To clean away the skin of nuts by scalding them with hot water.

Bleaching. A process for removing color from edible oils to obtain a more desirable light-colored oil. Diatomaceous earth, Fuller's earth, activated clay, and activated carbons are examples of bleaching agents. These are filtered out of oil before oil is consumed.

Blocking of powder. Compression of a powder in order to produce a nonflowing mass.

Bloom. Surface defect on chocolate candy formed either by a fat separation and resolidification or by sugar deposition.

Bob syrup. A cooked fondant. It is recrystallized by excessive seeding of the fondant mass used in the formula. It is an extender.

Boiling point. The temperature at which the vapor pressure of a liquid reaches atmospheric pressure.

Brake. Twin adjustable rollers used to reduce the thickness of confectionery product.

Brittle. A hard candy formulated with sugar, fat, salt, and nuts that is heated to "hard crack" stage.

°Brix. Measure of soluble solids in direct percentage by use of a refractometer, as percentage by weight of sucrose at known temperature.

Browning. Browning reactions are of two types, enzymic, which occurs on the cut surfaces of some fruits and vegetables, and nonenzymic (or Maillard reaction), which is a heat-induced condensation reaction between amino acids and sugars. The hydroxyl groups of sugars react with the amide groups of protein to form brown processed products which, in most cases, lower the nutritional and keeping qualities of thermally preserved foods affected by the reaction. Enzymatic browning is of little concern in candy manufacture.

Brown sugar. A crystallized "soft" mass of very fine sugar crystals covered with a film of refined cane molasses-flavored syrup. There are many kinds, usually designated as light brown, golden brown, and dark brown.

Btu (British Thermal Unit). The quantity of heat necessary to raise the temperature of 1 lb of water 1°F.

Buffer. A suitable mixture of acid and salt (or salt and base) that regulates or stabilizes the pH of a solution.

Candying. To preserve fruits by immersing them in syrup, thereby increasing their sugar content.

Calorie. A unit of energy. A kilocalorie is the amount of heat needed to raise 1000 g of water 1°C. This is the nutrient Calorie.

Caramel. A chewy candy, consisting of sugar, corn syrup, invert sugar, fats, milk, butter, and so on that have been heated to caramel consistency (244–252 F).

Carbohydrate. A polyhydroxyl derivative of a saturated aldehyde or ketone of structure CHO.

Catalyst. A substance that alters the rate of a chemical reaction but remains unchanged at the end of the reaction.

Celsius (°C). Celsius equals °F — 32 × $^5/_9$.

Cherries, Maraschino. Produced from sweet, pitted cherries (*Prunus avium*), bleached and colored with a certified, FDA-approved red dye.

Chelating agents. Substances such as citric acid, ethylenediamine tetraacetic acid (EDTA), ascorbyl palmitate, and phosphoric acid, which bind pro-oxidants such as Fe and Cu ions.

Chocolate. A semiplastic food derived from fermented cacao beans. The dried beans are roasted, dehulled, and degermed, and the remaining nib is milled. The resulting chocolate liquor is combined with sugar, with or without cocoa fat, flavors, lecithin, and so on to produce chocolate. It must comply with FDA Standards of Identity for Cacao Products to be called chocolate. Three types are bitter, milk, and sweet.

Chocolate liquor. A slurry derived from ground cacao nibs in which the fat is the continuous phase and nonfat solids are the suspended phase. Chocolate liquor contains about 54% cocoa butter.

Cocoa. The press cake of chocolate liquor is crushed and pulverized into a fine powder. There are three types: low fat, medium fat, and high fat.

Cocoa butter. Produced from chocolate liquor, usually by pressing out the soft fat, known as butter to the trade.

Cold test. A test used in oils to determine the time it takes for solids to appear in an oil when stored at 32°F (0°C). If an oil remains clear after 5.5 h at 32°F, it is considered properly winterized.

Compound coating. When fats other than cocoa butter are used in combination with cocoa, a coating is designated as a compound coating.

Conching. The final mixing process in the production of milk chocolate that is responsible for the development of flavor.

Colloids. Substances so very finely divided that surface forces become an important factor in determining their properties.

Concentration. The amount of a substance per unit volume.

Conjugated fats. Fats in which double bonds alternate with carbons of a single bond.

Consistency. A measure of firmness or resistance to deformation.

Cornstarch. Derived from wet milling of corn; made up of glucose units in polymer form. Cornstarch is a thickening agent, gelling at about 170°F (77°C). It is used to make starch jelly candies.

Corn syrup. Carbohydrate or syrupy liquid derived from wet milled corn containing specified degrees of sweetness and viscosity depending on dextrose equivalent (D.E.) and degree of hydrolysis. Glucose syrup (as it is also called) is a clarified, concentrated, aqueous solution of the products obtained by the incomplete hydrolysis of cornstarch. The solids of glucose syrup contain not less than 40% by weight of reducing sugars calculated as anhydrous dextrose.

Crack. One stage of heating a solution of sugar to 290°F to remove moisture content and effect the desired texture.

Creams. Candy in which the sugar has been crystallized into microscopic crystals.

D.E. Dextrose equivalent in corn syrup, expressed as the weight of reducing sugars contained therein. It is calculated as anhydrous dextrose. Syrups are classified from low to extra high conversion types depending on their dextrose content. The higher D.E. syrups normally have a thinner viscosity.

Dehydrating agent. A substance that extracts moisture from another substance, such as CaO, silica gel, or molecular sieves.

Deodorization. A process of removing pro-oxidant and odor-producing substances from a fat or oil.

Depositor. A machine that places candy in its liquid or semiplastic form in cornstarch impressions or molds.

Dextrin. A cooked carbohydrate starch that imparts a chewy texture to candy.

Dextrose. A monosaccharide known as glucose. Its solubility is lower than that of sugar, and it can crystallize and cause grittiness.

Dialysis. A process in which a semipermeable membrane is used to separate colloidal particles from substances in true solution.

Dilatometry. A technique that measures the proportion of solid and liquids present in a fat. It is extended over a range of temperatures and expressed as solid fat index (SFI). Solid fat index is sometimes called solid fraction index.

Dipping. Synonymous with coating of candy.

Distillate. A substance that has been vaporized and then condensed in a separate vessel.

Doctor. Cream of tartar, acids, or syrups used in candy to prevent crystallization or "sugaring."

Dusting. Coating confections with icing sugar, fine sugar crystals or antistick powders.

Dutch or alkali processed chocolate. Cacao nibs or cacao beans are processed in whole or in part with any optional alkali ingredient such as potassium carbonate.

Emulsifiers. Food-grade surfactants used in chocolate to retard "bloom," such as sorbitan monostearate and polyoxyethylene 20 sorbitan monostearate. Lecithin is a natural emulsifier used to thin chocolate before enrobing.

Emulsion. One liquid dispersed in another liquid, achieved by mechanical means or chemical or both. Butter is a water-in-oil (w/o) emulsion.

Encapsulated flavors. Flavors coated with gums to stabilize or give controlled-flavor release.

Engrossing. In panning, the building up of a sugar coat on centers.

Enrobing. To coat or dip chocolate. This is done under controlled conditions depending on the type of coating used.

Enzymes. Organic catalysts, a typical enzyme is invertase.

Equilibrium. A state in which a chemical reaction and the reverse reaction are taking place at the same time.

E.R.H. (Equilibrium Relative Humidity). Point at which product neither gains nor loses moisture.

Essential oils. Oils contained in most plant material, e.g., oil of orange, peppermint, and so on. These are used in flavoring applications.

Eutectic. A mixture of two or more substances producing a lower melting point than anyone of the ingredients alone.

Evaporation. The changing of a liquid to vapor by heat or reduced pressure or both.

Extract. A substance dissolved in ethanol such as flavors.

Extrusion. The general principle used in the extrusion of products is to feed the product into a hopper that distributes it over a set of fluted rollers. It is used especially for plastic products such as nougat, fudge, marzipan, and caramels.

Fahrenheit. Temperature scale on which 32°F is the freezing point of water and 212°F is the boiling point.

Fat bloom. Development of unstable fat crystals on the surface of chocolate.

Fats. Glyceryl esters of aliphatic monocarboxylic acids, known as triglycerides. They are solid at room temperature.

Fatty acid composition. The percentage of different fatty acids found to be present in characteristic or source triglycerides.

Feather. A stage in boiling sugar syrup.

Fermentation. A chemical reaction caused by organisms or enzymes or both.

Flow. Term used generally for flow behavior of a confection.

Fluidity. The reciprocal of viscosity.

Fo. Processing parameter; number of minutes a product is processed at 250°F.

Foam. Coalescing of air into fat.

Fondant. A mixture of sugar, corn syrup, invert sugar, and water cooked to 240°F, cooled to 115°F, and creamed on a Dayton-type mixer until it loses its gloss. This forms a supersaturated solution.

Fourrés. A confection with a soft center of nut paste, marzipan, or preserved fruit dipped in fondant crème—the true bonbon.

Formula. Same as "recipe" expressed in parts by weight in lieu of volume.

Formers. Series of horizontal rollers arranged to form boiled syrup into a thin rope.

Fractional distillation. A process of separating by distillation two or more liquids that have different boiling points.

Frappé. An aerated mixture of corn syrup and invert sugar containing albumen or egg whites or soy protein whipping agents.

Fructose. Monosaccharide reducing sugar.

Fudging. Unplanned crystallization of sugar in a confection.

Fungi. Microscopic vegetative cells that can sporulate and grow on candy surfaces when the temperature and humidity conditions are right. High humidity can cause condensation of moisture on the surfaces of candy removed from the cooler environs, especially if the candy is unwrapped or is poorly sealed or the wrapper is permeable to moisture.

Gelatin. Gelatin is a gelling agent produced from the bones and skin of animals. It is classified according to bloom strength.

Glacéed fruits. Fruit slices or pieces sweetened with sugar and corn syrup.

Glycerin. A syrupy liquid used as a humectant in candy ($CH_2OHCHOHCH_2OH$).

Graining. Unplanned crystallization of sugar in a confection.

Gums. Thickening or gelling agents used in candy production such as CMC, Methocel®, or gum arabic. Gum candies can be thickened with cornstarch and gum arabic.

Hard ball (248 F). When a drop of hot syrup is added to chilled water, it forms a rigid, pliable ball when placed between the fingers.

Hard butter. A processed fat characterized by a high solid fat index up through 80°F (27°C) with a narrow melting range and brittle texture and melting at body heat.

Hard crack. A stage in the boiling of sugar (290°F).

Heat of vaporization. The amount of heat required to change 1 g of liquid into vapor at the same temperature.

High fructose syrup. A sweet-tasting syrup assaying about 50% glucose and 42% fructose on a dry weight basis. It contains only 71% solids. If a higher solids syrup of this composition were produced, there could be a glucose crystallization problem. It can be produced by reacting glucose syrup with isomerase at 60°C.

Honey. A sweet extract produced by honey bees in a comb from specific nectars or mixed floral sources which give it a distinct flavor. It is separated from the comb by mechanical means and usually is heat treated to reduce crystallization or graining. On the average, honey is about 18% water, 34% dextrose, 41% levulose, 2% sucrose, 1% dextrins and gums, and 4% of such varied items as acids, pollen grains, beeswax, pigments, proteins, enzymes, aromatic bodies, and higher alcohols.

Humectant. Used in candy to retain moisture; lowers the E.R.H. to reduce drying out.

Humidity. The amount of water vapor per unit of gas.

Humidity, relative. The ratio of the amount of water vapor in the atmosphere to the amount necessary for saturation at the same temperature.

Hydrogenation. A process in which hydrogen is added to certain unsaturated fats and oils in the presence of a catalyst, giving them a higher melting point and stability.

Hydrolysis. A reaction between a salt and water in which an acid or base is formed.

Hydrolytic rancidity. Glycerides that are hydrolyzed to glycerol and constituent fatty acids by microbes, enzymes, alkali, or combinations thereof. Lauric fatty acids in coconut oils are especially prone to hydrolytic rancidity or "soapines" if not processed properly.

Hygroscopic. A substance that takes up moisture, but does not dissolve in it.

Indicator. A substance that changes color at a definite pH, such as phenolphthalein.

Interfacial tension. The force operating at the boundary between two liquids, not soluble in each other.

Inversion. A process by which sucrose is converted into glucose and fructose by hydrolysis.

Invertase. An enzyme derived from yeast to control texture, prevent fermentation, and prolong the shelf life of cream and cordial centers. A commercial product is called Convertit®.

Iodine value (IV). An expression of the degree of unsaturation of a fat. Normally the stability increases as iodine value decreases. Hydrogenation lowers the iodine value.

Ionization. The splitting up of a molecule into ions of opposite electrical signs. For NaCl, the cation = Na^{+1} and the anion = Cl^{-1}.

Isomers. Molecules having the same number and kinds of atoms arranged differently within the molecule, giving them quite different properties.

Jelly. A semiplastic consistency obtained when pectin is added to sugar at a controlled pH.

Joule. In electrical energy, a watt second.

Kilo. Metric prefix for 1000.

Kilowatt. One thousant watts, expressed as kW.

Lakes. Dyes or pigments absorbed on a substrate of alumina; insoluble in water and color by dispersion.

Latent heat. The heat absorbed or liberated in changing a mole of substance from one state to another at a fixed temperature.

Leaching. Transfer of sugar or starch from one confectionery type to another.

Leavening agent. A product that releases carbon dioxide that aerates a product.

Lecithin. A phosphorylated natural emulsifier, normally derived from soybeans.

Lemon oil. Essential oil expressed from a lemon. There is a fivefold concentrate. There are synthetics which emulate or duplicate their natural counterpart.

Levulose. Fructose.

Lime oil. Essential oil expressed from a lime; these come in synthetic flavors, and one oil is noted as Mexican terpeneless.

Linoleic acid. A fatty acid with 18 carbons and 2 cis double bonds. It is an essential fatty acid in the human diet.

Linolenic acid. A naturally occurring fatty acid of 18 carbons and three cis double bonds.

Liquid sugar. A solution of sucrose in water of about 67% solids.

Lovibond color. A color rating of a fat or oil on evaluation with a Lovibond

Tintometer, e.g., 1.5R — 15Y (R = red, Y = yellow).

Low-calorie food. Not more than 40 Calories per serving (see Reduced calorie.)

Mannitol. A sugar alcohol; $CH_2OH(CHOH)_4CH_2OH$.

Marshmallow. An aerated food product composed of various sugars, water, gelatin, and a flavor. It may contain an edible food color and a humectant.

Marzipan. An almond paste-type confection containing sugar, frappé, water, fat, and milk powder.

Massecuite. Mixture of boiled syrup and sugar crystals.

Mazetta. Egg albumen, sugar, and corn syrup heated and whipped into a frappé. It is used in cream center candies.

Mogul. A machine synchronized and timed to produce formed candy.

Molar solution. A solution that contains one mole of solute per liter.

Molasses. A heavy-bodied liquid derived from sugar cane refining of about 80% solids. It has a distinct flavor, with the U.S. Grade C possessing a stronger flavor than Grade A. It exhibits acidity and thus is a natural leavening agent. It contains invert sugar and minerals and may be purchased as either sulfured or unsulfured.

Monoglyceride. A chemical compound composed of one glycerol unit. It can result from combination of fatty acid and glycerol or from the breakdown of di- and triglycerides.

Monostearates. Emulsifiers.

Nibs. During fermentation of raw cacao beans, the cotyledon gains moisture and changes to a fissured structure. When dried, it breaks into pieces called nibs.

Nougat. An aerated confection produced from a supersaturated sugar solution base. There are nongrained, grained, and light nougats. To make a nougat lighter, more egg albumen is normally added.

Off-flavor. A flavor that is objectionable or atypical of the product. For example, a "cardboardy" flavor is not considered desirable in a milk powder. Rancidity or staleness are considered off-flavors.

Orange oil. Natural oil, usually cold pressed or extracted from the flavedo of the orange. Oils can be concentrated or folded to higher strengths.

Orange peel. Peel from an orange diced into sections and preserved with sugar and corn syrup.

Osmosis. The passage of a liquid or gas through a semipermeable membrane from a denser to a less dense concentration.

Oxidation. A process in which oxygen combines with another substance.

Sometimes termed autoxidation, it is a reaction of oxygen with double bonds in unsaturated fatty acids. Ordinarily, it is a slow process at low temperature.

Panning. Coating of confections by the revolving pan method. In Europe the name "dragee" is applied to these coated confections. Three types are soft coated, such as jelly beans; hard coated, such as sugar-coated chewing gum; and chocolate coated, such as "bridge mix."

Pectin jellies. Candies that are thickened or gelled with pectin. Other essential ingredients are water, sugar, and acid.

Penuche. Fudge made with brown sugar. Also known as panocha.

Phase. The homogeneous part of a system.

pH. Negative logarithm (base 10) of the hydrogen ion concentration. A pH of 7 is neutral while any value less than 7 is considered acidic and a value more than 7 is alkaline. The pH range is from 1 to 14.

pK. Point at which the acid and the base are in equal concentration.

Pregelatinized starch. A starch that disperses in water and thickens without heating.

Plastic creams. A so-called hand-rolled cream or plastic cream whose texture is such that it is capable of holding its shape without support.

Polymerization. The formation of long chain byproducts or gums when a product is heated. Fats are much more subject to polymerization when bonds are in the conjugated position.

Polymorphism. The ability to exist in two or more crystalline forms, such as cocoa butter in a chocolate coating.

Polyunsaturated. Fatty acids that contain two or more cis methylene interrupted double bonds per fatty acid chain, such as linoleic acid.

Preserves. Fruits treated with sugar and corn syrup or glacéed.

Press cake. Compressed cocoa powder after removal of cocoa butter.

Pro-oxidant. A substance that speeds up a reaction of fats with oxygen; for example, Fe and Cu ions may be encountered in commercial candy products. Chelating agents such as citric acid, EDTA, and ascorbyl palmitate can be effective in binding these minerals.

Prosthetic. The non-amino part of a conjugated or joined protein.

P/S ratio. A ratio of polyunsaturated fatty acids to saturated fatty acid components in a fat.

Pulling. Repeated folding of a high-boiling sugar syrup. Carried out on the revolving arms of a pulling machine.

Quench. Cool; quenching is sometimes used after roasting coffee.

Rancidity. A characteristic off-flavor development that can occur in candies.

Three types recognized are oxidative, ketonic, and hydrolytic rancidity.

Reduced calorie. Food containing at least one-third fewer calories than its full-calorie counterpart (see Low calorie).

Reducing agent. A substance that removes oxygen from another substance.

Refractometer. An instrument that can measure soluble solids by refracted light, expressed in either refractive index values or as direct percentage of sucrose.

Reversion. An undesirable change in flavor of a refined fat or oil, particularly in soybean oil, which contains some linolenic acid. The reversion flavor is described as "grassy," hay-like, bitter, or the like.

Roasting. Heating a product with dry heat for a specified time and temperature.

Salt. A compound containing a metal or base radical and an acid radical, e.g., NaCl.

Sanding. A term used in candy terminology to designate placing a coating of sugar around gums and jellies.

Saponification. The hydrolysis of fats and oils by an alkali.

Saturated solution. A solution in which the solute in solution is in equilibrium with the undissolved solute.

Seeding chocolate. The adding of β-crystals of cocoa butter into chocolate to temper or stabilize it before enrobing.

Shelf life. Serviceable life of a product.

Sizing. Spinning of candy such as salt water taffy to a predetermined size before cutting and wrapping.

Soft ball (238 F). When a drop of hot syrup is added to chilled water, it forms a soft ball that flattens when placed between the fingers.

Solute. The substance dissolved in a solvent.

Solvent. The constituent in a solution which is present in the largest amount.

Sorbitol. A sugar alcohol used in candy as a humectant. It is sold in crystals or in liquid form at a solids content of about 70%.

Specific gravity. The ratio of the weight of a unit volume of a substance to the weight of the same volume of water.

Specific heat. The heat required to raise the temperature of 1 g of a substance 1°C.

Stability. The relative resistance of a product to change or breakdown.

Starch. Polymers of dextrose that can be either branched or straight-chained. There are hundreds of different commercial starches; each is designed for a particular application.

Stearine. A fat product that has been hydrogenated until it is hard and brittle.

Sugar. Sucrose, the most concentrated ingredient in candy.

Sugar bloom. Development of fine sugar clusters on the surface of chocolate.

Supersaturated solution. One in which more solute is in solution than is present in a saturated solution of the same substances at the same temperature and pressure. Most candies fall into this category.

Surfactant. A synthetic detergent (syndet) cleansing agent. When it is added to water, it lowers the surface tension of the soil, which is released into the water by dilution. Syndets are either anionic, cationic, or nonionic surface active agents. Very few anionic surfactants are commercially available.

Surface tension. The contractive force of a surface.

Taffy. A type of chewy or semihard candy such as salt water taffy.

Temperature. That condition which determines whether heat will flow from one body to another.

Tempering. The procedure used to stabilize chocolate into its most crystalline form.

Tenox®. Registered tradename of the Eastman Kodak Company's antioxidants.

Tincture. A solution of alcohol.

Tribasic acid. An acid containing three replaceable hydrogen atoms per molecule.

Unsaturation. A property of a fat or an oil that contains less hydrogen than a saturated fat (designated by double bonds).

Valence. The number of atoms of hydrogen (or its equivalent) that will combine with or be replaced by the atom or radical in question.

Vanilla. A natural flavor derived from orchid pods. It has been replaced to a large extent by ethyl vanillin, its synthetic counterpart. Vanilla is one of the most popular flavors used in candy.

Vapor pressure. The (partial) pressure exerted by a vapor.

Viscosity. The apparent resistance to flow of a liquid; sometimes measured in centipoises.

Vitamin. An organic chemical substance found in many different foods and essential for bodily health and vigor.

Water (H_2O). Perhaps the most important ingredient used in candy production. It is now regulated by the Environmental Protection Agency National Interim Primary Drinking Water Regulations as covered by FR Vol. 40, No. 248, dated December 24, 1975. The regulations took effect June 24, 1977. FDA regulates water in the plant.

Weak acid. Acids that are only slightly ionized in water solutions.

Wrapping materials. Any material either natural or synthetic used to wrap candy.

Xylitol. A 5-carbon sugar alcohol used as a sweetener in chewing gum.

Yeast. Microscopic cells that reproduce by budding and are osmophilic, thus capable of infecting candy.

Zein. Corn protein; it is used in specialty candy coatings, as in panning.

APPENDIX

Water to be Added to 100 Gallons of 67.5 Brix (36.3 Baumé) Syrup to Obtain Various Baumé Syrups

Brix	Baumé Desired	Water to Add Gal	Water to Add Lb	Final Gallons Syrup
67.5	36.3	0	0	100.0
67.0	36.0	0.9	7.5	100.9
64.9	35.0	5.1	42.5	105.1
63.0	34.0	9.2	76.6	109.2
61.0	33.0	13.9	115.8	113.9
59.1	32.0	18.7	155.8	118.7
57.1	31.0	24.0	199.9	124.0
55.2	30.0	29.3	244.0	129.3
53.3	29.0	35.1	292.4	135.1
51.4	28.0	41.4	344.9	141.4
49.5	27.0	48.1	400.7	148.1
47.6	26.0	55.4	461.5	155.4
45.7	25.0	63.2	526.5	163.2
43.8	24.0	71.7	597.3	171.7
41.9	23.0	81.0	674.7	181.0
40.1	22.0	90.5	753.9	190.5
38.2	21.0	101.7	847.2	201.7
36.3	20.0	113.8	948.0	213.8

Source: Crompton & Knowles

Safe Densities for Storage of Invert Syrups in Possible Presence of Seed Crystals

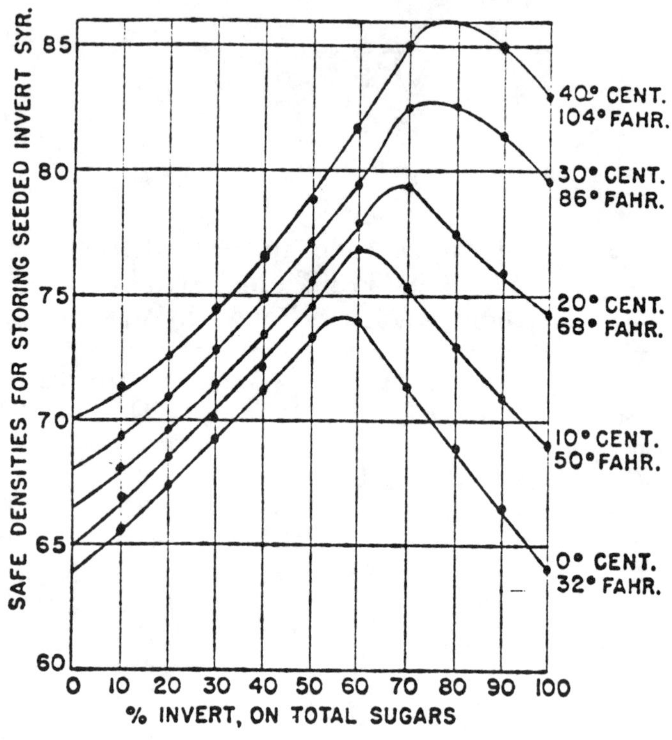

Curves are based on assumed average supersaturation factors of 1.0 for sucrose and 1.75 for invert sugar.

Source: C and H Sugar Co.

Physical Data

The Pearson Square

(a rapid calculation for diluting or concentrating syrups)

Problem: You have 500 lb of No. 11 **Nulomoline** testing 50% sugar solids. How much dry sugar must you add to the syrup to bring the sugar solids up to 72%?

Calculation: Draw a rectangle with diagonals and insert figures.

A is % solids in syrup
B is % solids in dry sugar
C is % solids in finished syrup
D is the difference between B and C
E is the difference between A and C

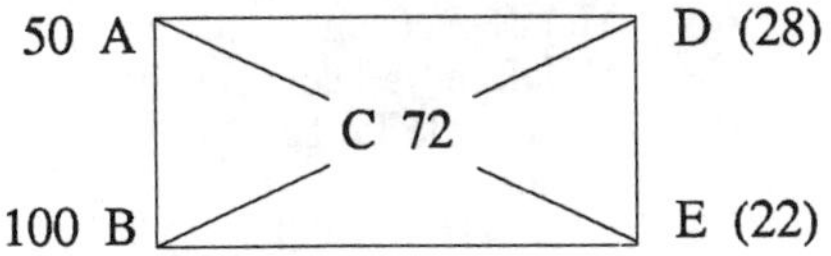

$$D = 100 - 72 = 28$$
$$E = 72 - 50 = 22$$

$$\frac{500\text{ lb}}{D} \times E = \frac{500}{28} \times 22 = 393 \text{ lb of dry sugar}$$

Total weight of syrup with 72% solids is 893 lb.

Problem: You have 1000 lb of regular **Nulomoline** testing 77% solids. How much water must you add to reduce the syrup to 70% solids?

Calculation:

A is % solids in syrup (77%)
B is % solids in water (0%)
C is % solids in finished syrup (70%)

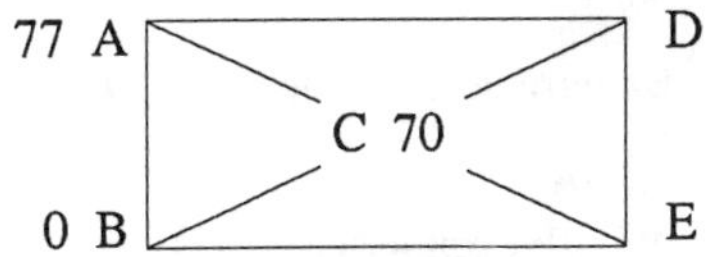

$$D = 70 - 0 = 70$$
$$E = 77 - 70 = 7$$

Then $\frac{1000}{D} \times E = \frac{1000}{70} \times 7 = 100$ lb of water

The total weight of regular **Nulomoline** with 70% solids is 1100 lb.

Source: Crompton & Knowles

INDEX